JOHN DIVINE

# BITCOIN OPTIONS

## Energy Storage, Risk Management, Real Yield

JOHN JAMES DIVINE

Photo Cover Image by J. Borba front, back cover, spine, and interior book design and setting by estrategy. tech.

This book is written by **John James Divine.**

To book the author, contact: connect@estrategy.tech

## IN MEMORY OF DARRIN JOSS.

*Darrin was my mentor and taught me the foundation for nearly everything I have learned in markets.*

# DISCLAIMER

Any views expressed herein by John Divine are those solely of the author and do not represent the view of the author's employer or any person(s)or group(s) the author is or is not affiliated with. The information herein is taken from the author's personal experience, and sources believed to be reliable. Information presented in this book is intended for informational and educational purposes only and is not guaranteed by John Divine or any of the author's affiliates as to its accuracy or completeness, nor are any trading results that may be mentioned. This educational content does not constitute a recommendation to buy or sell options contracts, futures contracts, swaps, or securities. It does not constitute trading advice or a solicitation of financial services. Under no circumstance will John Divine, his employer, or any persons or groups the author has an affiliation with be liable for any injury or damages resulting from the use of the information provided by John Divine's books. The data is presented "as is" without warranties to the fullest extent permissible to applicable laws. To the extent the information provided links to or cites from third-party content that John Divine does not create, John Divine does not endorse those sites or citations in any way, and John is not responsible for the content of such other sites or citations.

# EDITORIAL FOREWORD

In the dynamic world of financial markets, innovation unlocks new possibilities. John Divine's "Bitcoin Options: Energy Storage, Risk Management, Real Yield" is a comprehensive guide to understanding the intricate intersections of bitcoin and the world of options trading. As a digital asset, bitcoin has evolved into a revolutionary force in finance, offering a decentralized alternative that challenges systemic inequalities.

Humans cannot navigate the terrain of bitcoin and its options without addressing the broader context of monetary systems and civil society structures. John's narrative presents bitcoin as a decentralized alternative that is a bulwark against systemic disparities.

John explores bitcoin as a store of value and a catalyst for reshaping the perception of money, energy, and risk. This book is an invitation to delve into the transformative potential of bitcoin and options trading, urging readers to embrace next-generation financial instruments in the hyper-connected age for humanity.

Towards the future,

**M. Lulu Mayorga**

Editor

# AUTHOR BIOGRAPHY

John James Divine is an author, speaker, market maker, liquidity provider, and financial market professional focused on structured products, derivatives, lending, borrowing, and spot market execution for digital assets. John began his career as a market maker for Chicago Mercantile Exchange and Intercontinental Exchange listed derivatives, specializing in precious metals, industrial metals, interest rates, energies, and forward volatility curves. John has also operated within environmental products, exchange-listed, and over-the-counter (OTC) for compliance and voluntary participants.

John moved into the digital asset arena trading as an execution specialist for an industry-leading hedge fund before joining BlockFills. He currently engages with institutional participants in digital asset markets on crypto strategy development and execution. John Divine earned his Bachelor of Science in Renewable Energy Technology from Illinois State University and has a Master's Degree in Energy Systems and Fuels Management from Southern Illinois University, Carbondale.

# ACKNOWLEDGEMENTS

I extend my deepest gratitude to the pillars of support who have enriched my journey—my cherished family, steadfast friends, insightful teachers, dedicated teammates, invaluable colleagues, and esteemed partners. A special acknowledgment to Nick Hammer, Gordon Wallace, Perry Parker, Neil Van Huis, Adam Krauszer, and the entire Blockfills team for their unwavering collaboration and shared commitment.

I am indebted to the belief and encouragement offered by my editor, Lulu Mayorga, at eStrategy.tech. Your faith in my work has been a driving force in bringing this project to fruition. My heartfelt thanks to Kyle Johnson, Jen Bawden, John Slazas, James Putra, Tone Vays, Dan Schak, Paul Sacks, Ari Pine, Eddy Sfeir, Spencer Riegle, Taylor Ortiz, Mike Ortiz, Chris Mack, Ronnie Virissimo, Nasser Mohsin—your contributions in time and energy have been invaluable along my journey.

To those with whom I have had the privilege to collaborate in trading and execution, I express my sincere gratitude. This journey has been shaped by the collective efforts of an exceptional network, and I am profoundly thankful for all of you.

**John Divine**

Chicago, Illinois | 2024

# INTRODUCTION

## Practical Optionality

Understanding how options contracts can provide solutions to yield objectives and risk management goals is a tool for people who utilize the Bitcoin protocol and the associated digital asset, *bitcoin*, for energy storage and transfer from work that has yielded excess, *unspent energy*. This book will reference pricing models, historical data, volatility, and put/call ratios but will not attempt to teach the layers of theoretical mathematics that go into such models – as these arrangements of inputs provide outputs that seek to mimic reality but are not exactly reality. Instead, **this book will focus on practical applications that utilize option contracts for bitcoin holders seeking ways to optimize store-of-value asset management.**

# TABLE OF CONTENTS

CHAPTER

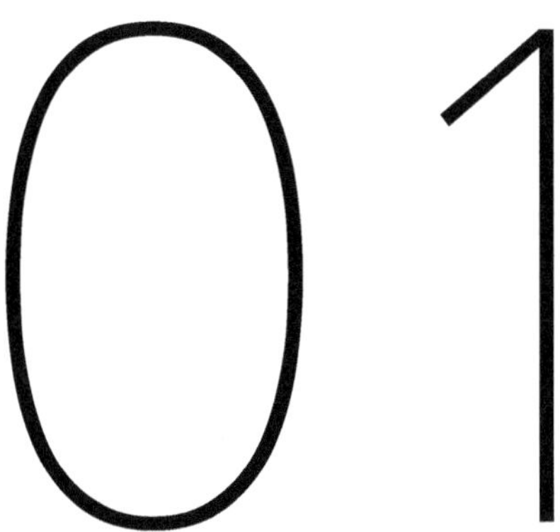

# MARKETS ARE MADE FOR HEDGERS

## MARKETS ARE MADE FOR HEDGERS

As with most financial or technological innovation stories, there is a preamble, a seed, a beginning that would be engraved into history, adding another block in the timechain of existence. In the story of options markets, the plot is based on the premise of risk transfer between producers and consumers of goods and services in society. The interconnectivity of people thrives on the ability to exchange ideas, innovations, natural resources, refined products, and services. Within this layer of personal interaction and unity, regional and cultural diversity strengthens societies and propels the growth curve exponentially with the free flow of energy.

The critical variable to the entirely interconnected interpersonal system is ***trust.*** Trust is the root of connection, an unconditional bond reverberating harmony between all people. With *trust*, there is *risk*; the risk that trust can be upended, either intentionally or unintentionally. Actual (historical) and potential (implied) systemic trust failure sets the foundation for price discovery and volatility in commodities and other tradable assets. Through international trade routes, these assets spawn from people's time, energy, and effort. For this, the ability to manage risks and profit from them has fascinated and intrigued people over thousands of sun cycles.

From the mathematical side of the brain, applied through finance, engineering, and economics, formulas for pricing risk and modeling for future expected price ranges of goods and services were made real. Analytics arose from the brain's creative side and was applied through philosophy and psychology. This thinking focused on behavior, cycles, symbolism, astronomy, and seasonality. With advances in technology, the input of computing power can rapidly provide an output for analytics, formulas, and models. Thus, the stage has been set at the crossroads for modern risk management strategy and execution.

The potential for systemic trust failure can genuinely facilitate risk transfer between producers and consumers. The flow of energy in the trading of goods and services from producers to consumers is driven by a risk and reward analysis, propelling the willingness of people to partake in the flow of life itself. Technological, and financial innovation in risk management, from forward agreements and futures contracts to options and structured products, seeks to mitigate risks within the realm of *known knowns*, *known unknowns,* and the acknowledgment of *unknown unknowns* through the practical application of risk modeling and micro/macroeconomic analysis.

## HEDGING TIME DECAY

Time is life's most precious gift, providing the potential for energy accumulation, storage, and transfer. Whether running, reading a book, meditating, praying, designing a building, writing code, going to the store, drilling for water, giving a speech, creating music, operating a machine, capturing steam, sowing seeds, milking cows, or herding cattle; *time* enables the *opportunity* and *experience* of energy transfer relative to *capacity* for energy storage and transfer.

Remembering high school physics class, energy is *the ability to do work*. Therefore, energy transfer is doing work against an object or objective. The laws of thermodynamics state that:

1. Energy is stored; it is neither created nor destroyed; it is only transferred.

2. Entropy, or the randomness of an isolated system, always increases.

3. Entropy: The entropy of an isolated system approaches zero as the temperature approaches absolute zero.

Elaborating on the 2$^{nd}$ and 3$^{rd}$ laws of thermodynamics, energy transfers from high to low-temperature areas, and heat transfers thermal energy between systems due to a temperature difference. The foundation of energy storage and transfer is rooted in the principle that an object has potential energy or *stored energy* when it is not in motion. Once a force has been applied to an object, the potential energy changes to kinetic energy or *energy of motion*. Observing the first law of thermodynamics, if energy is the ability to do work and it can neither be created nor destroyed, only stored, and transferred, **then** the capacity for a person to store and transfer energy is directly related to their ability to do work. The ability to work is individualistic or can be assignable through energy transfer or *money*.

People go to work to make a living. They transfer their energy at work toward supplying goods or services that other people demand in exchange for predefined energy compensation at an agreed-upon quantity. The salary or wage of compensation represents potential or unspent energy. Thus, people experience time as an opportunity to do work. In some cases, a person's work expenditure yields a salary or wage that can be accumulated and stored as potential energy available for transfer in the future.

What then shall people of this world agree upon as a store of energy mechanism, or medium of exchange, for goods and services? In the barter system, people traded goods and services for other goods and services without a medium of exchange. Through time, the advent of energy stores came to fruition - from rice and wheat, which could be stored underground, to gems, oils, incense, salt, precious stones, and eventually precious metals, of which people found enough value through rarity and awe, that they would be willing to exchange their time, energy, and effort for these elements.

Eventually, lords and kings involved themselves with stores of value, issuing mints of coins derived from precious metals and stones to represent stored energy as a medium of exchange that could be traded with other tribes and kingdoms in exchange for goods and services. The fiat paper currency systems people live with today evolved from these early mediums of exchange systems based on gold and silver coins. With centralized power structures, such as with lords and kings, the temptation of greed is often a downfall. Fiat paper currency is nothing but a product of greed.

When people began to earn precious metals as salary or wage, they stored their earnings through self-custody or in storage vaults hosted by *trust* companies beholden to the local empire. Early trust companies begat banks who invented loaning stored energy for interest payments against a percentage of precious metal deposits. With greed permeating further, banks invented paper notes redeemable for the precious metals in the vaults that backed them. It was much easier to carry and transfer paper notes between people than to carry the actual precious metal around, so the system flourished into the dominant mechanism for trade. Eventually, the standard medium of exchange between people for energy transfer was adopted and standardized. Paper notes became *a promise* that precious metals could be retrieved by the note holder at the vaults storing the metals anytime.

It's easy to see where the story goes from here. Banks that issued redeemable paper notes soon realized that there were *almost always* more precious metals in their vaults than people coming in to redeem the physical metals. Ingenious plans were devised to print more redeemable paper notes than precious metals readily available in the vaults for redemption, ushering in the age of money backed by nothing but promises of repayment, otherwise known as *debt.*

# MEDIUM OF EXCHANGE

Fast forward to present times, the current situation in our monetary systems can be roughly traced to the sequence of events described previously. Banks have cornered the market for defining and creating the acceptable mediums of energy exchange known today through a well-designed network of central, regional, and TOO BIG TO FAIL institutions. The chosen medium to centralize financial power has been leveraged by fiat currencies. Pieces of paper backed by promises of repayment or retribution doled out to the masses through labor and debt agreements for shelter, education, transportation, and revolving credit.

The mechanics of money, a medium of energy exchange that people trade their time, energy, and effort for, should be addressed as a subject matter that is a *known known*. However, the mechanics of money are *an unknown unknown,* meaning, most people in society do not know what money is. *This knowledge gap is driven* primarily by central banks' radical takeover of money that transformed historical store value assets into a *debt asset* known as *fiat currency*. To create a new fiat currency, debt must be issued. Debt comes with a cost known as *interest*. Interest payments add further pressure to the social system as they compound through time as a cost of capital. Remembering high school physics class, energy cannot be created nor destroyed; it is only transferred. Fiat currency creation accompanied by debt interest implies energy creation from simply lending a medium of energy exchange backed by the promise of repayment rather than providing added value through goods and service offerings.

**The fiat currency system bends the laws of physics against ordinary people.** Thus, the most impactful question humanity must ask in the 21st century is: "**What will people agree upon for a medium of energy exchange, or money, to store and transfer energy between each other as interconnectivity exponentially expands through time?**" And what assets will be viewed as stores of value that can stand the *test of time* known to present ample opportunities for greed, fraud, theft, and censorship? Will it be land? Will it be gold? Will it be hydrocarbons? Food? Art or automobile collections? A digital asset of electrons from a decentralized, open source, and cryptographically secured protocol that proves ownership through heat transfer? When humans offer energy transfer or WORK, what shall be considered an accepted form of compensation in exchange for the relative value of energy transfer itself?

With macroeconomic questions on store-of-value energy storage sustainability- i.e., which assets will store energy over time within societal structures? The question often leads to more questions rather than direct answers. This predicament led to the *advent of risk management.* Portfolio asset diversification emerged at the forefront of responsible asset management for people who sought to plan for long-term, even generational accumulation of stored energy available for transfer.

When dealing with risk management techniques, the core of the practice is understanding KNOWNS. Within our human experience on earth, there are *known knowns* or things a person knows that they learn through trial and error. There are *unknown knowns,* things a person knows but doesn't know they know–like reflexivity or survival instincts. There are *unknown unknowns* or things that a person doesn't know, and they don't know that they don't know, and *known unknowns are* things a person knows they don't know.

K*nown knowns,* known *unknowns, unknown knowns,* and the acknowledgment of *unknown unknowns* are the center of risk management implementation for asset managers and those seeking to store and accumulate intergenerational energy. The ability of a person to grasp *known knowledge* allows for strategy development and innovation around portfolio allocations and the trading of goods and services.

*Known unknowns* can be considered scenarios involving too many variables to quantify an exact solution, such as: What will the exchange rate be for store-of-value assets like gold and bitcoin relative to the dominant mechanism for international trade at some point in the future? Or what would the cost of food and energy sources like wheat, rice, gasoline, diesel, and electricity be relative to the dominant medium of exchange or the going rate for housing and commercial real estate?

*Acknowledging unknowns* furthers the possibility of deploying risk management strategies for preserving stored energy. Often acquired from stories of other people's experiences, *unknown unknowns* are rooted in highly disordered environments like power vacuums, war, famine, natural disasters, or the effects of changes in local, regional, and global precipitation or temperature, either naturally or unnaturally.

## HEDGING TYRANNY

Before we get into practical applications, *why* are we talking about bitcoin in the first place? The *Bitcoin protocol* is the technology layer upon which the digital asset *bitcoin* exists as a digital representation of an unspent transaction, known as a UTXO. In a physical sense, bitcoin is the most basic form of electrical energy. With a lowercase "b," bitcoin originated through electrically charged neuron membranes from Satoshi Nakamoto's mind at the crossroads between internet technology and cryptographically secured computation powered by the same electrically energized spark as a lightning bolt.

On January 3rd, 2009, amid the Great Financial Crisis, the first bitcoin was mined through verifiable *Proof of Work* from computational effort. This allowed a peer-to-peer transaction of unspent energy to be processed and settled without a third-party intermediary. During this time, a movement known as Occupy Wall Street manifested amongst the 99% of people who were angry at the top 1% of the wealthiest population.

Wealth can be defined as an individual's capacity to store and transfer energy. In the Occupy Wall Street movement, the protesters asserted that 99% of Americans were suffering, having once again been robbed and looted through the rigging of financial markets, interest rates, perpetual war, restricted rights, and freedoms, devalued savings accounts through inflation, and millions in unemployment lines. Living in a dejected and crestfallen economic system, people seemed to agree that banks and those controlling all the "money" were corrupted beyond repair. Something must be done to stop them, as the Occupy Wall Street movement slogan signaled: **"THE BANKS GOT BAILED OUT, WE GOT SOLD OUT!"**

# BITCOIN

The Bitcoin protocol realigns "money" to its true nature- - verifiable energy transfer. It aims to reshape how people build societal structures and interact through local and international commodity and foreign exchange markets. With a foundation of decentralized computing power running open-source software that elegantly solves the Byzantine Generals Problem, the Bitcoin protocol and the Proof of Work consensus mechanism that defends system integrity have provided a solution to those seeking to store and transfer accumulated energy without exposure to third-party permissions and fundamental metric manipulation that affects relative valuation.

The Great Financial Crisis confirmed to the technological generation born in the *New Millennium* that banks and governments could misuse their monetary authority to inflate asset prices and stagnate working wages by sending savings accounts on a one-way path to zero. The 1% deploys this strategy to ensure 99% of the global population will be left fighting amongst each other for resources and jobs as divide and conquer techniques to maintain control are achieved through fiscal domination. Bitcoin is a call option against the greed of centralized, tyrannical authority.

The goal of Bitcoin is simple: defend it from tyrannical forces that control people's energy transfer mechanisms and mediums of exchange through the centralized creation and control of "money." To safeguard energy transfer, Bitcoin ushered in a decentralized digital asset, ***bitcoin***, controlled by no one and accessible to everyone. Bitcoin is a way to store personal energy as a savings account and transfer that energy to others as electronic cash. The critical economic difference between this store of value - bitcoin, and the traditionally accepted fiat currencies of old, the USD, CNY, RUB, EUR, INR, GBP, JPY, AUD, RND, BRL, etc., is the **finite supply of 21 million bitcoin**. This limited supply of bitcoin is defined by the protocol and backstopped by networked computational energy that defends a system that must prove work to validate bitcoin transactions.

Central banking and governmental fiat currencies have infinite currency supply and are susceptible to greed and inflation through the printing of fiat currency. In comparison, the Bitcoin Protocol remains incorruptible through hash rate inputs consistently reaching consensus truth of transactional data from peer-to-peer bitcoin energy transfers. Although you can interact with whole or fractions of bitcoin, this finite supply and censorship resistance is a method of energy exchange without needing third-party gatekeepers. This is where bitcoin gains its initial value, and it is solidified by open-source architecture and a robust network that grows stronger the more it is attacked by corrupt forces seeking to fight against a systemic change that can transform the broken store of energy standards through public awareness and public accessibility.

The U.S. dollar is the world's reserve fiat currency in international trade. This means international contracts for trading goods and services are settled in USD. If Saudi Arabia wants to sell oil, they receive U.S.A. dollars as a settlement for exchanging barrels. Does China want to buy wheat? The transaction for the wheat exchange with China is paid in USD. The country that enjoys having its local money as the world reserve fiat currency benefits tremendously from the natural demand for its energy transfer mechanism, U.S.A. dollars. The responsibility of controlling the world's reserve currency is rooted in the trust of the global population. Unity, innovation, and prosperity must be driving forces by decision-makers rather than the doom of greed, control, and manipulation.

We could look back thousands of years and identify the dominant mechanism for facilitating personal energy transfer, although that would take us way off the scope of this book. Since 1450, there have been six significant world reserve currency periods. Portugal (1450–1530), Spain (1530–1640), Netherlands (1640–1720), France (1720–1815), Great Britain (1815–1920s), and from 1945 to present day, the United States of America. The system of reserve currency is supported by centralized authority over currency creation. Adding new currency influences the distribution of natural resource extraction and refinement technology, deployment of military tactics, and the management of strategic shipping routes and receiving terminals for international trade of goods and services.

The newfound global interconnectedness achieved through evolution from the *mechanical to the electrical energy age* will cause money to experience a reimagination from a re-examination by international consensus. What will the agreed-upon energy transfer mechanism, or "money," between willing buyers and sellers worldwide be? In the 21st century, we have vast historical data to support the thesis that centralized control over money creates a concentration of wealth and an inevitably unfair monetary system based on a hierarchy that did not develop from the people's will. With an unelected hierarchy that drives consequential decision-making on energy transfer flows, society rewards status and greed while disincentivizing productivity and equality.

# DEFINING UNELECTED HIERARCHY

Born from congressional order on December 23rd, 1913, the latest iteration of centralized control over currency in the United States was made possible by enacting the Federal Reserve Act. The 63rd United States Congress pushed the legislation through voting based on proposed solutions for providing a backstop to the rampant bank runs caused by a lack of trust between U.S.A. dollar depositors who used banks to keep their stored time, energy, and effort safe and the people running those energy storage institutions. The 99% were often forced to "run on the bank" and withdraw their U.S.A. dollars based on rumors or facts of fraud and mismanagement at their energy bank of choice. As banks often face the temptation of leaving customer energy deposits alone or gambling with them in commodity or debt markets, *trust* has become a rare reality between bank operators and their customers, even at the most well-respected institutions.

The Federal Reserve Act set out to create a central energy bank that could lend currency to other energy banks and the people seeking to store and transfer energy themselves - citizens (government by the people, for the people) in times of need. This private corporation would go on to become the most powerful entity in the world, even succeeding in debasing U.S.A. dollars from a gold-backed standard (temporarily in 1933 and permanently in 1971) to a promissory note – one in which the citizens of the United States, or anyone with a social security number), along with all the natural resources within the nation's borders, would become collateral for a debt-based U.S.A. dollar. This debt-based dollar is a promise of repayment in the future, at interest, to be paid to the lending institution during the present.

U.S.A. debt dollars are Federal Reserve central bank notes. There is a reason the word TRUST is in focus when you examine the piece of paper. TRUST with the aggregate of your life's work by holding your stored energy in paper notes with an infinite supply, which the central bank determines through the setting of interest rates unveiled to the public at regularly scheduled Federal

Open Market Committee (FOMC) meetings. TRUST that everything is going to be OK. TRUST that centralized powers will not make mistakes or be susceptible to influence from tactics to increase their power and the public dependence on the Federal Reserve. Ambiguity is favorable for the system; seemingly hidden behind a veil is the mechanics of something so critical for people to function in society as it has evolved - control over the legally accepted personal energy storage and transfer mechanism used to buy groceries, pay for rent, and settle local, state, and federal taxes. TRUST that your savings account will not be eroded by interest payments and inflationary forces from increasing the supply of sanctioned paper currency for the masses to store their energy in, along with the ever-increasing tax burden born from the former.

# TRUST, BUT VERIFY

To better understand how bitcoin is a risk mitigation solution for storing and transferring unspent energy and why risk management techniques against the world reserve fiat currency for bitcoin holders will be a significant factor in wealth preservation in the coming decades, we can explore datasets distributed by the Federal Reserve itself.

Below is a graph of Total Public Debt incurred by the United States government through the issuance of treasury bonds bought primarily by the Federal Reserve central bank with newly created U.S.A. dollars, along with select corporations, hedge funds, and other foreign governments and central banks.

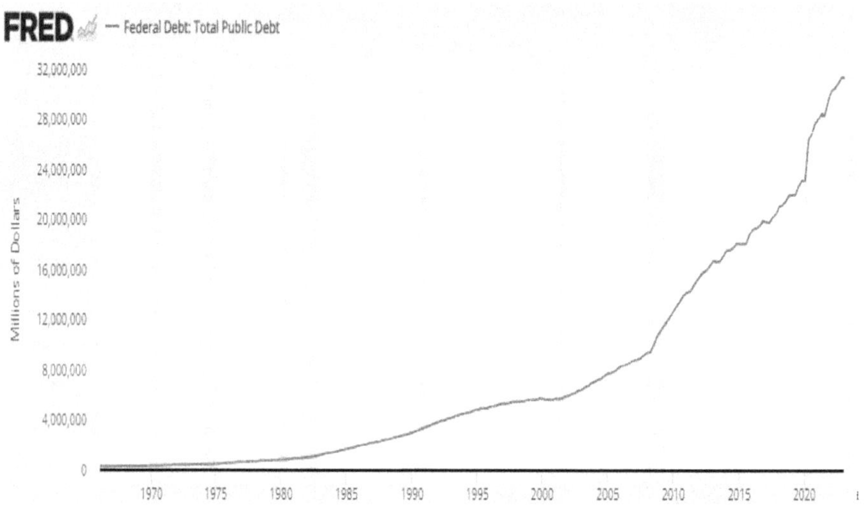


The following chart illustrates the amount of interest the U.S.A. government must pay to service this debt, generated through tax revenues placed on salaries and wages along with any asset sale, seizure, or confiscation from citizens and residents.

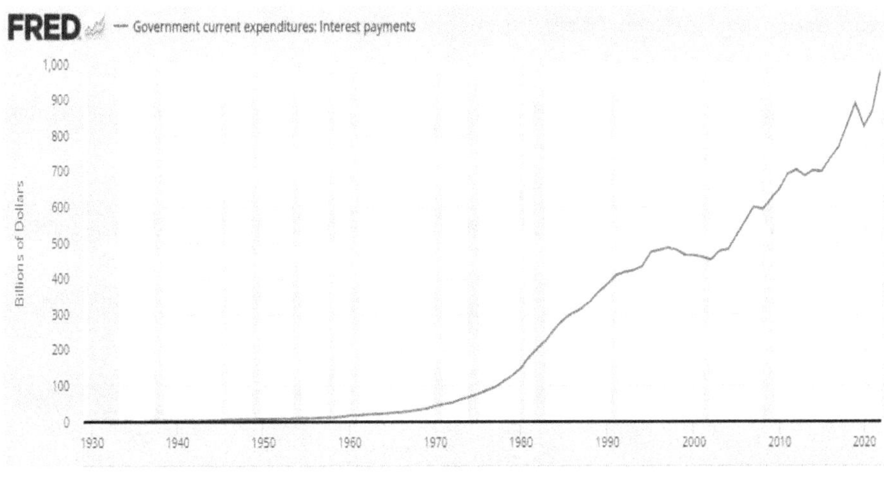


ii

Notice the data point to the right of the Government Interest Expenditures chart. That point is mathematically guaranteed to move higher if we continue to cover our eyes from the outright theft of current and future generations through debt burdens perpetuated by those controlling currency. Even as we hit the USD one TRILLION-dollar annual interest expense obligation incurred by the federal government and passed on to the U.S.A. taxpayer, there is no end to the expanding liability.

The first question that often comes to mind after reviewing these charts is how the tax–paying population agrees to live and work in a system that places future burdens on the ability to generate and store energy. How does a societal structure built on the principle of a democratic republic, where the people vote for peers who will sit in positions of power over all others, elect leaders that would mismanage taxpayer liabilities to such a degree that the slope of the charts for total debt and interest expenditures looks like the drawing of a massive mountain with a peak that has yet to be sighted?

One potential answer stems from the wage mirage. Although people are making more U.S.A. dollars per hour of work, which would seem like common sense given the productivity gains of people from the expansion of education, technology, and global interconnectivity, the actual REAL PURCHASING POWER of that increased amount of U.S.A. dollars per hour people are receiving in exchange for their time, energy and effort has barely budged since the 1960s:

**Americans' paychecks are bigger than 40 years ago, but their purchasing power has hardly budged**

*Average hourly wages in the U.S., seasonally adjusted*

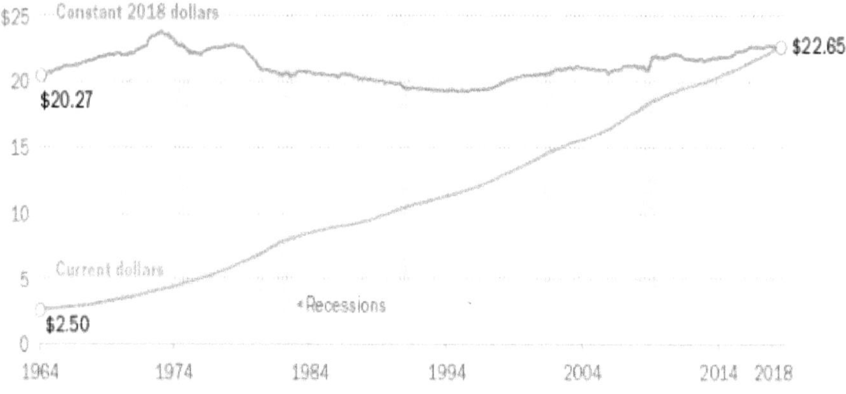


Note: Data for wages of production and non-supervisory employees on private non-farm payrolls. "Constant 2018 dollars" describes wages adjusted for inflation. "Current dollars" describes wages reported in the value of the currency when received. "Purchasing power" refers to the amount of goods or services that can be bought per unit of currency.
Source: U.S. Bureau of Labor Statistics.

PEW RESEARCH CENTER

iii

The following chart (M2) demonstrates the USD in market circulation, commonly referred to as the *money printer chart.* It visually depicts the creation of U.S.A. dollars, distributed to the system through the issuance of debt and the issuance of interest on all debt outstanding. This chart shows why hourly wages are up, but purchasing power is flat despite productivity gains, and why taxpayer interest expenditures are rapidly expanding.

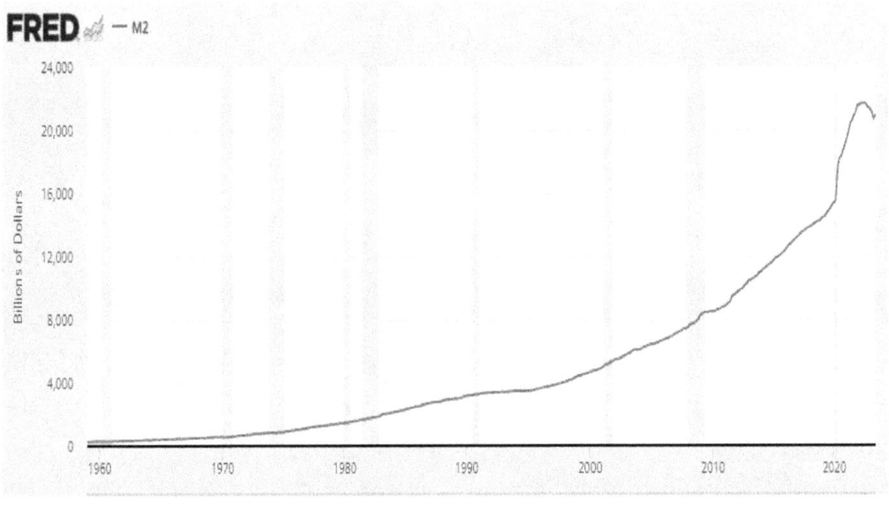


iv

Who wins in the game of currency printing? Data from the FED reveals the answer. The following chart shows the percent share of total net worth held by the **top 1%**, fast approaching 33% of all wealth in the United States.

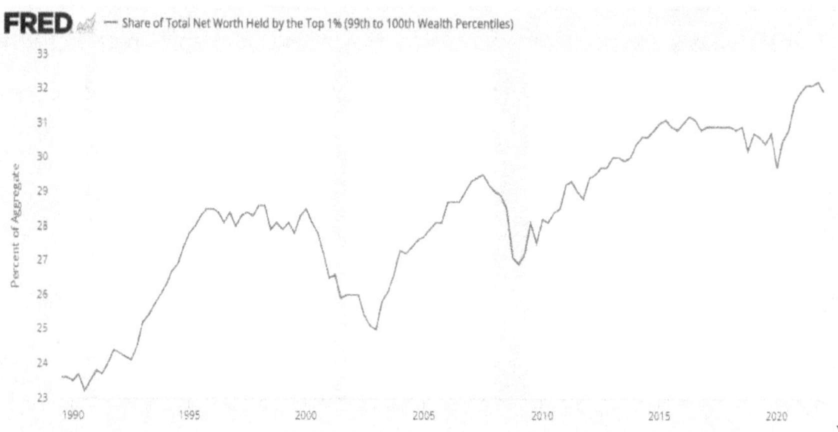

v

Below, you can observe the total net worth held by the **top 0.1%.** The Central Bank System transfers wealth from the working class to the aristocracy.

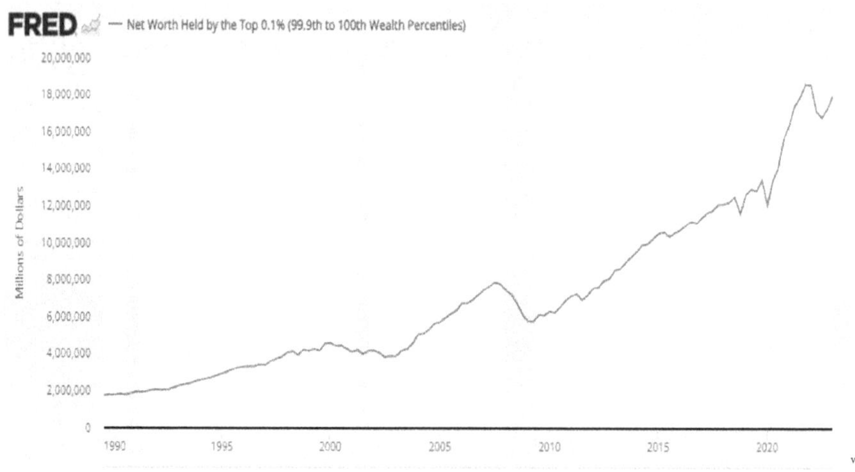

vi

# HEDGING SAVINGS ACCOUNTS

Hopefully, this brief introduction on "why bitcoin" has alerted, confirmed, or supported your thesis that U.S.A. dollars or any fiat currency controlled by a centralized authority is no place to save your time, energy, and effort accrued through the transfer of your energy (work).

Below, we will continue with chart analysis, although we will pivot to traditional stores of value-priced against the current world reserve currency, U.S.A. dollars.

Since the Federal government decided to debase the U.S.A. dollar from a gold-backed standard in 1971, accompanied by immense influence from central currency planners, the exchange rate for gold has been on an upward trajectory, escalated by outlandish debt accumulation from government expenditures worldwide.

*Gold (USD / Oz):*

vii

To understand gold's market capitalization or net USD conversion value, we can quantify the total above-ground gold in storage or circulation and multiply that quantity by the U.S.A. dollar exchange rate. Unlike bitcoin, where the supply is fixed, gold has a variable supply, recently estimated at 205,238 metric tons above ground by the World Gold Council in 2021. This means that the net USD value of gold in circulation is around **$13 trillion**[viii].

Silver, like gold, is another precious metal that has held a historical store of value, deriving from coins minted and circulated by past civilizations as currency or jewelry and industrial applications. For the total supply of above-ground silver, data from the US Geological Survey shows there are over 1.5 million tons above ground, giving Silver a market cap of roughly **$1.1 Trillion** based on conversion prices of $24.00 per ounce[ix].

*Silver (USD / Oz)*

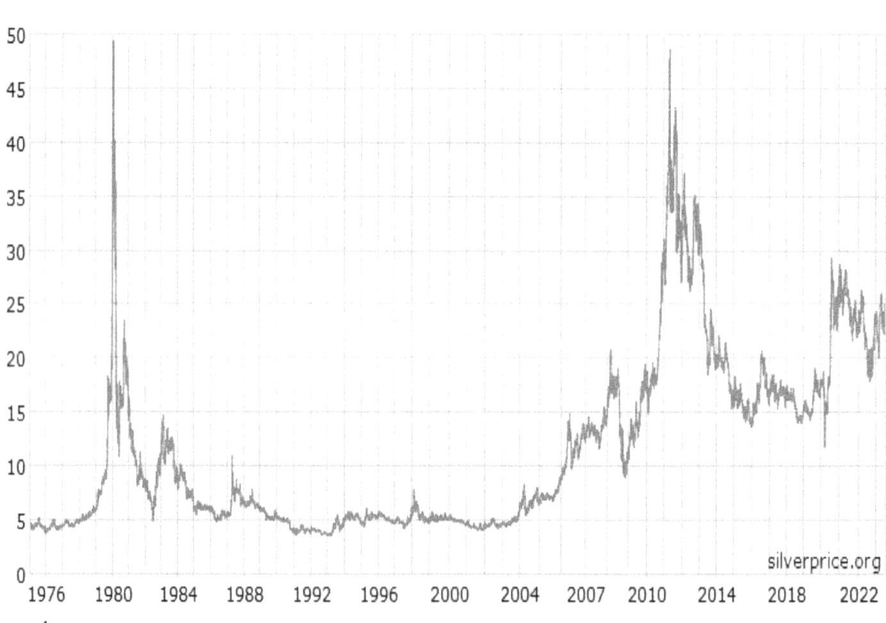


[x]

For bitcoin supply and market capitalization analysis, we do not need to rely on a government agency or private organization that incentivizes research on behalf of clients. With bitcoin, the supply is hardwired into the Bitcoin protocol. It is defined as **21 million bitcoin**, with a distribution schedule for block rewards of bitcoin that are released to bitcoin "miners."

Miners validate transactions on the blockchain by inputting computational energy into a Proof of Work hashing algorithm, adding a new block of transactional data to the Bitcoin Blockchain. This *proof of work* secures the integrity of the network in a distributed manner. Below is a chart illustrating total bitcoin block reward distributions:

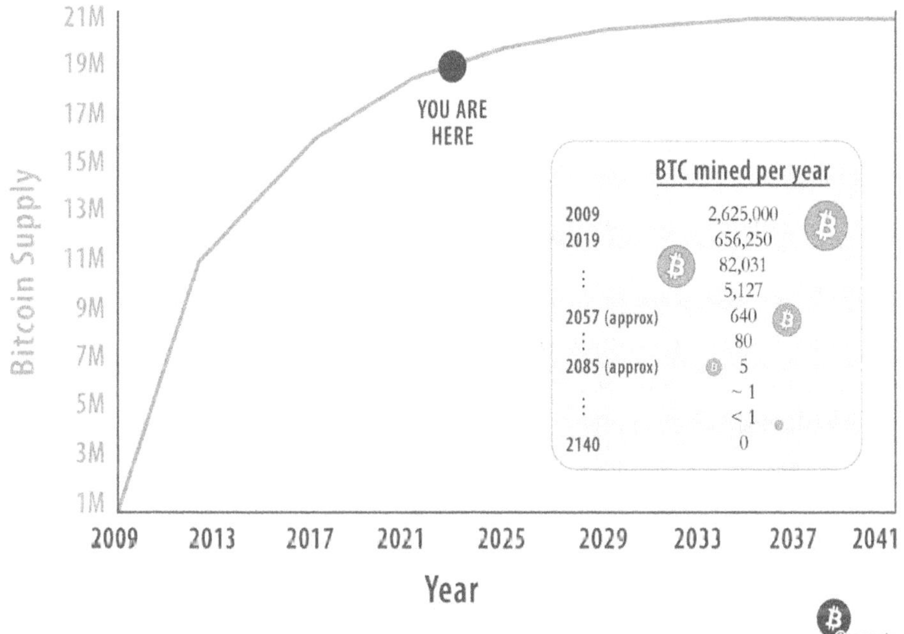


xi

Compare the previous bitcoin total supply chart to the U.S.A. dollar total supply (M2) chart below.

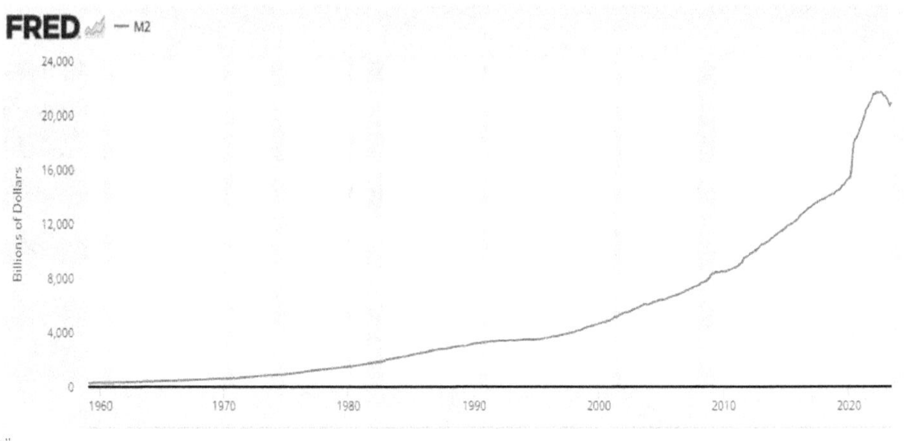


xii

As defined in the Bitcoin protocol, bitcoin block rewards experience a "halving" event every four years, during which bitcoin distributions to miners are halved until no more bitcoin can be distributed through block rewards. Using back-napkin math, by the year 2140, the bitcoin network will be supported through transaction fees alone, as block rewards will have reached zero.

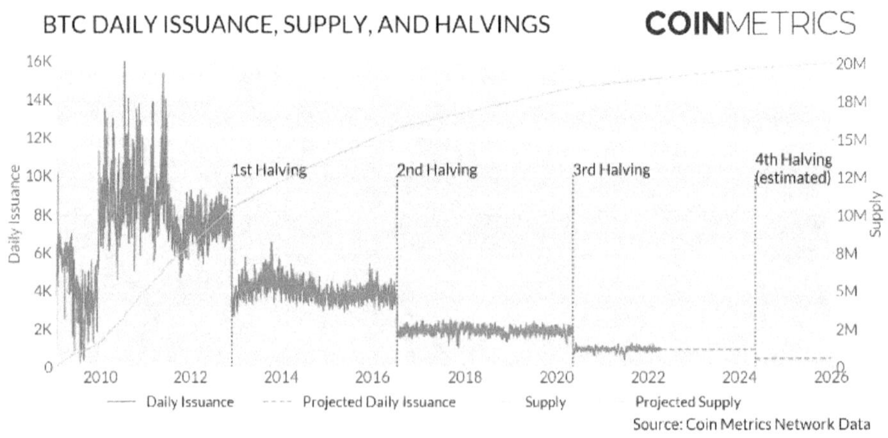


xiii

Given the transparent fundamentals of bitcoin supply, we can calculate the total market capitalization in U.S.A. dollars by taking the total supply that will ever exist, 21 million, and multiplying by the current USD exchange rate for one bitcoin, which is $29,500 at the time of writing this book. Thus, the total bitcoin market cap at the time of writing is **$696 billion.**

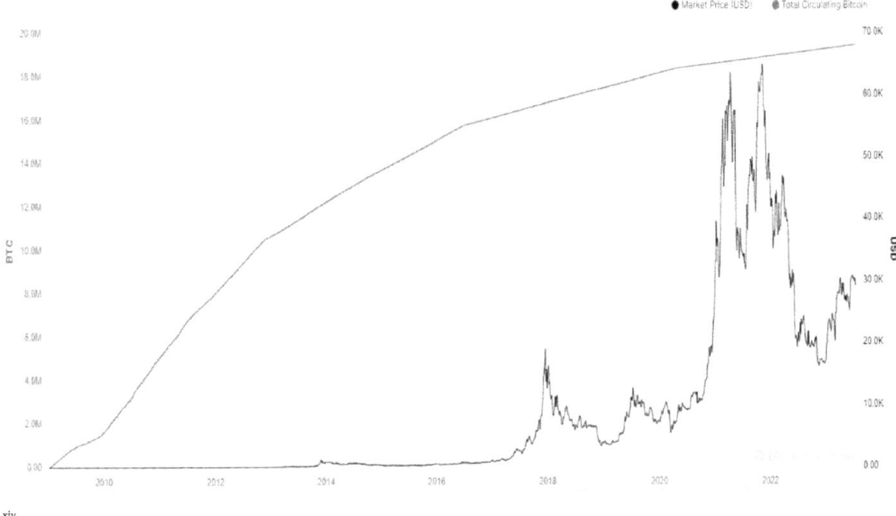


xiv

## HEDGING TOWARD BITCOIN

If you traverse the internet deep enough, options history research is a well-documented subject that has found consensus on Aristotle's origination story. Aristotle first described a written account of a CALL option transaction in his book *Politics*. In this genesis block of data, written around 332 BC, Aristotle tells of Thales of Miletus, sometimes called the Father of Science, who lived in Ancient Greece from around 624 – 545 BC.

Thales was a mathematician, philosopher, astronomer, and statesman. As a member of the seven sages, he is considered the first known Greek Philosopher, notably gracing our collective consciousness with the words "Know Thyself." Despite this imprint on human history, in his times, Thales was considered a poor man when measured against the local mediums of exchange. Thales began experiencing critique from people in Miletus, questioning how a man of such intelligence could have such little MONEY. To Thales, the pursuit of money held very little interest, a distraction to his purpose of pursuing science and the mind's inner workings.

In Miletus, vineyards produced a bountiful commodity – olives. To derive a range of refined goods from olives, olive press operators played a vital role in processing the annual harvest. Rooted in the function of supply and demand, the cost to utilize an olive press operation increases during times of exceptional yield. When harvest yields are modest or worse, olive presses' cost, use, and operation decrease. Therefore, olive press operators, the producers of olives, and those who consume the refined product face annual price risk due to the historical and implied volatility of seasonal harvests.

Thales, using his knowledge of astronomy, is said by Aristotle to have approached regional olive press operators ahead of a season he anticipated to yield exceptionally. He negotiated contracts with all the olive press operators, where in return for placing a deposit, Thales would have the *right*, but not the *obligation*, to hire the olive presses at a fixed cost. This contract allowed the olive press operators to generate revenue by selling the option contract to Thales. It defined a price with acceptable profit margins where Thales could exercise his right to utilize those presses as stated in the contract.

With the subsequent season's arrival, Thales' speculation of exceptional yield came to fruition, and the demand for olive presses far exceeded the supply. Thales could access the olive presses for the predefined price he negotiated with the olive press operators in his region. That optionality allowed Thales to profit from the high demand for olive presses. Thales' speculation rendered him wealthy as measured by the local medium of exchange. Thales' motivation was not rooted in the pursuit of profit but in promoting the appreciation of study and acquiring knowledge; the former critics were silenced. His message reverberated throughout the region and through the *timechain of existence*. Though Thale's story occurred long ago, his ideas persisted and made themselves into this book.

**Below is the English translation of Aristotle's writings on *Politics*:**

[1259a] [1] "And Apollodorus1 of Lemnos have written about both agriculture and fruit-farming and similarly others also on other topics, so these subjects may be studied from these authors by anybody concerned to do so; but in addition, a collection ought also to be made 2 of the scattered accounts of methods that have brought success in business to certain individuals. All these methods are serviceable for those who value wealth-getting, for example, the plan of Thales3 of Miletus, which is a device for the business of getting wealth, but which, though it is attributed to him because of his wisdom, is really of universal application. Thales, so the story goes, because of his poverty, was taunted by the uselessness of philosophy, but from his knowledge of astronomy, he had observed while it was still winter that there was going to be a large crop of olives, so he raised a small sum of money and paid round deposits for the whole of the olive-presses in Miletus and Chios, which he hired at a low rent as nobody was running him up; and when the season arrived, there was a sudden demand for several presses at the same time, and by letting them out on what terms he liked he realized a large sum of money, so proving that it is easy for philosophers to be rich if they choose, but this is not what they care about. Thales then is reported to have thus displayed his wisdom, but as [20] a matter of fact, this device of taking an opportunity to secure a monopoly is a universal principle of business; hence, even some states have recourse to this plan as a method of raising revenue when short of funds: they introduce a monopoly of marketable goods. There was a man in Sicily who used a sum of money deposited with him to buy up all the iron from the iron mines, and afterward, when the dealers came from the trading centers, he was the only seller. However, he did not greatly raise the price, but all the same, he made a profit of a hundred talents4 on his capital of fifty. When Dionysius5 knew of it, he ordered the man to take his

money with him but clear out of Syracuse on the spot,6 since he invented means of profit detrimental to the tyrant's affairs. Yet this device is the same as the discovery of Thales, for both men contrived to secure themselves a monopoly. An acquaintance with these devices is also serviceable for statesmen, for many states need financial aid and modes of revenue like those described, just as a household may, but to a greater degree; hence, some statesmen even devote their political activity exclusively to finance. And since, as we saw,7 the science of household management has three divisions, one the relation of master to slave, of which we have spoken before,8 one the paternal relation, and the third the conjugal9—for it is a part of the household science to rule over wife and children (over both as over freemen, yet not with the same mode of government, but over the wife to exercise republican government and over the children monarchical[xv])."

## TIMECHAIN

In Aristotle's writings, we uncover the genesis block in the story of options contracts. With the first *call option* transaction ever recorded, Thales bought the right, but not the obligation, to own olive presses at a predetermined price (strike price) during a predetermined time (expiration date) for an agreed-upon deposit cost from buying the contract (premium).

Although the name "Call Option" did not exist while Thales was alive, *call options* are options contracts that grant the call option owner THE RIGHT, but not an obligation, TO BUY an underlying asset at a negotiated price during a predetermined time. A PUT option has the inverse optionality, meaning the owner of a PUT option has THE RIGHT, not an obligation, TO SELL the underlying asset at a negotiated price during a predetermined time.

For every buyer of a call or put option granting the rights to optionality, there is a seller of that optionality, who assumes the obligation to sell or buy the underlying asset from the option buyer if the buyer exercises their rights to the call or put option.

Fast forward two thousand years from ancient Greece to 16th and 17th century Europe, where people continued to deal with the ever-present laws of supply and demand, quality control, and settlement assurances. These persistent issues were the foundation for innovation in forward agreement and option contract marketplaces. The port town of Antwerp, Belgium, created a commodities exchange in 1531 that allowed the trading of "to arrive" contracts. As a port town, feedstock producers, consumers, and speculators traded in contracts for commodities that had yet to arrive in the port, allowing for the deployment of risk mitigation techniques. Whale oil, grains, salt, fish, and poultry were the leading markets[xvi].

After measurable success in Antwerp, London unveiled a similar exchange in 1571, the Royal Exchange. With liquidity building in "to arrive" and other creative financial contracts, early versions of tradable instruments that resemble modern options contracts were created and issued on these exchanges and "over the counter" marketplaces. Speculators provided liquidity provision services to actual end users of the underlying products that were transferring price risk in these markets as a risk mitigation technique for their core businesses.

Following Antwerp and London, the Amsterdam Bourse was founded in 1611 and quickly became prominent. The Amsterdam Bourse marked a pivotal point in our historical reflection on options contracts as the exchange standardized forward agreements, eventually leading to options contracts that had listed expiration dates for both calls and puts. These call-and-put options focused on speculation around Dutch East India Company stock certificates. The earliest known options contract written in English is an agreement between Sir Bazill Firebrass of Mark Lane, dated July 29, 1687, to deliver 1,000 East India stock at 200 to Sir Thomas Davill on or before March 1, 1688, in return for a premium of 150 guineas[xvii].

The notorious Dutch tulip bulb frenzy between 1633 and 1637 saw rampant speculation in the price of tulip bulb commodities, famously crashing down and leaving many in financial ruin. Out of the rubble, it was found that much of the speculation came from forward agreements. The relevance of option contracts to the mania arises from the legal outcomes related to the collapse of prices from the peak, usually traced to February 3, 1637.

Upon thorough review by the courts at the time, it was noted that many of the sellers of the forward agreements did not have possession (the ability to sell) tulip bulbs when the contract was signed, thus leaving the contract holder with the optionality to refuse delivery because the seller did not have possession of the commodity when selling the forward contract. Optionality was recognized in recourse as a risk mitigation technique for buyers of worthless contracts from sellers who did not have the means to deliver obligations when the contract was signed.

On the other side of the world, members of the Japanese elite, including the Samurai, were dealing with a commodities trade dilemma requiring risk mitigation techniques. The Japanese economy was centered around rice production and consumption, exposing the people to violent price swings, either up or down, that could lead to disastrous second-order shock waves throughout society. Often used as currency, the rice commodity market was primed for innovation during a slump in rice spot market prices.

As warriors, philosophers, and scholars, Samurai were compensated with rice. Being paid with rice means a price risk exists between the value of the local currency that could be redeemed from the sale of rice. This entanglement of commodity-based compensation led to the demand for the ability to lock forward prices for rice to be delivered at a specific date in the future. The Dojima Rice Exchange opened for business in 1697, providing a public, transparent venue for trading spot market rice and futures contracts that settled against the rice spot market. From standardization came more liquidity and price stability, along with the ability of the Samurai warriors, nobles, and other workers to manage the sale price of their rice compensation more efficiently.

## MODERN-DAY OPTIONS

During the 18th and 19th centuries, trade continued to build on past innovations in the prominent commercial hubs of Europe, although, in London, options trading was made illegal from 1733 to 1860. Eventually, the prominence of options trading reached the United States. Often cited as a pioneer of USA options, Russel Sage is said to have brought OTC options trading to the frontier by offering loans against stocks where he would combine the underlying asset with put options to create lending scenarios. Just as in England, US-based exchanges banned Options trading in the late 19th century until the - emergence of a pathway to government-sanctioned markets for options contracts with the Chicago Board of Options Exchange (CBOE) opening in 1973.

The city of Chicago, Illinois, plays a leading role in the origins of modern financial instruments used to manage risk in commodity and foreign exchange markets from the emergence of two critical exchanges: The Chicago Mercantile Exchange, which is still to this day the largest futures marketplace in the world, and the Chicago Board of Options exchange. The CBOE was opened in 1973, coinciding with professors Fisher Black, Myron Scholes, and Robert C. Merton's papers describing options pricing models based on various variables.

The formula described in *The Pricing of Options and Corporate Liabilities*, published in the Journal of Political Economy, would be named in infamy as the *Black-Scholes* model[xviii]. With mathematics providing a tailwind, this foundational model led to an injection of liquidity into a still obscure financial instrument. Myron Scholes and Robert C. Merton would win a Nobel Prize for their work. Despite being unable to receive this prize posthumously, Fisher Black is at the cornerstone of risk management and liquidity provision in global markets.

## MAKING A MARKET

For efficient marketplaces to emerge, three critical problems must be solved:

1. Contract Standardization
2. Liquidity
3. Settlement Procedures

Contract standardization defines the quantity of the underlying good or service. It sets terms for transactions on the contract within the marketplace, including time, location, delivery method, currency payable, and product quality. Liquidity is the depth of the market or the quantity of willing bids and offers for a specific contract. With more willing buyers and sellers, the potential for transparent price stability and supply and demand equilibrium increases. As liquidity builds, the ability for both end users and speculators to enter and exit positions significantly improves, furthering market confidence and incentivizing market participants to continue interacting within the marketplace of listed assets & instruments.

In addition to standardization & liquidity, the *confidence* of the market participants within the marketplace and the instruments traded also hinges on the assurance of settlement or ***trust*** in the functionality of the marketplace. Buyers and sellers have been and will remain hyper-focused on the ability to take delivery and be compensated for delivering goods and services. These three pillars of support to the marketplace: standardization, liquidity, and settlement, led to the migration of commodity traders to dedicated exchanges and to direct market access liquidity providers that enabled structural assurances for clearing trades and ensuring settlement.

When writing this book, bitcoin options can still be considered the bleeding edge for commodity derivative markets. The Bitcoin protocol (Jan 2009) was designed for decentralized peer-to-peer electronic cash transactions. The New Liberty Exchange is thought to be the first bilateral OTC venue for bitcoin transactions settled with USD via PayPal (Dec 2009).

Bitcoinmarket.com (2010) is considered the first public exchange, helping propel bitcoin across the $1.00 mark by 2011[xix]. The adoption curve was amplified by third-party centralized exchanges such as Kraken (2011), Bitstamp (2011), and Coinbase (2012) seeking to solve for user experience, liquidity & storage infrastructure. With the advent of derivatives exchanges, most notably BitMEX (2014), Deribit (2016), and Binance (2017), liquidity in forward contracts and options began to build. Innovation was motivating for participants at these early venues where accounts could be funded directly with bitcoin deposits and listed futures & options contracts that were settled with bitcoin.

The depth of market expansion and liquidity provision accelerated on September 17th, 2015, with the Commodity Futures Trading Commission (CFTC) ruling bitcoin as a *commodity*, not a security. On December 10th & 17th, 2017, the Chicago Board of Options Exchange and the Chicago Mercantile Exchange began listing bitcoin futures contracts settled into US dollars. By January 2020, the CME began listing bitcoin options contracts, which were settled in US dollars.

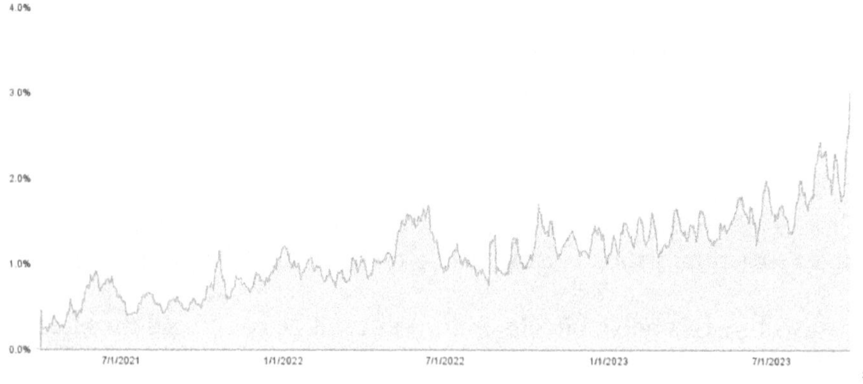


[xx]

# BORDERLESS BITCOIN

Regarding bitcoin prices, the most common reference rate is BTC/USD. King dollar holds firm even within the realm of bitcoin when facilitating the functionality of peer-to-peer energy transfer, as USD hegemony still firmly controls the world. Merchants can accept bitcoin but will often convert it to USD at the point of sale. When talking about Bitcoin more broadly, we are still at the stage of adoption where the relative value to USD is seemingly more important than exposing the US dollar for what it is – **debt.**

bitcoin is borderless electronic cash, controlled by no central entity and open to everyone. During the adoption phase of bitcoin, it is essential to remember the relative value between bitcoin and the global currency market. From The Brazilian Real, Japanese Yen, or Chinese Yuan to European Euros, Russian Rubles, Indian Rupee, and Nigerian Naira, in many cases, bitcoin prices have remained elevated despite depressed price action in BTC/USD. In notably mismanaged currency markets, from the printing of fiat currency and subsequent inflationary forces, bitcoin reference prices are near, at, or exceeding all-time highs as shown in the following charts.

[xxi]

xxii

Observing the charts for bitcoin priced in Argentine Pesos and Turkish Lira, the mismanagement of monetary policy is evident relative to bitcoin's sound money principles. The central banks governing these fiat currencies have effectively lost control of the relative value between their fiat currency and that of *sound money*.

With bitcoin, people who live in countries that experience currency mismanagement from centralized authorities now have an *unconfiscatable*, immutable, borderless, and censorship-resistant electronic cash that they can use to store or transfer their accumulated energy.

As the United States dollar remains perched upon the throne of settlement instruments for international trade of goods and services, other fiat currencies that are mismanaged lose value against USD, bitcoin, precious metals, industrial metals, oil, natural gas, timber, grains, fruits, vegetables, fibers, livestock, poultry, and refined products. Before the all-encompassing international adoption of bitcoin, multiple central bank failures will likely occur across the legacy currency markets. Given this, the US dollar will have a chance to continue its façade of value for longer than it should, albeit at the expense of non-US dollar earners, which is most of the global population. - The world's people need to wake up and open their eyes to the outright theft of their time, energy, and effort through the paper fiat currency debt system. Civilization can experience a chance to unite amidst an open-source energy transfer mechanism for the trading and settlement of goods and services globally. Why? To propel people further through the timechain of existence, reverberating as harmony from the *sound money* principles of bitcoin and a transparent governance hierarchy.

xxiii

CHAPTER

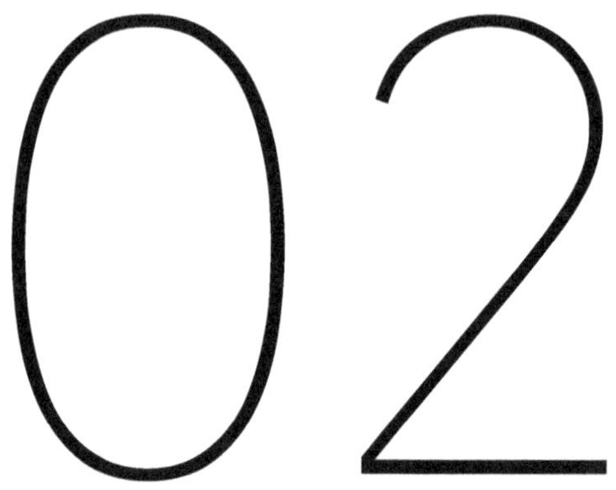

# BITCOIN OPTIONS FUNDAMENTALS

# CALLS & PUTS

There are two types of vanilla options contracts: *calls & puts*. For every option contract, there is a buyer and a seller. A *buyer* of an option is said to be *long* the contract, while a seller of an option is short the contract.

Call options give the owner the *right* (not the obligation) to buy an underlying asset at an agreed-upon future date at a predetermined price. *Sellers of call options receive an upfront premium payment from the call buyer* in exchange for the *obligation to sell* the underlying asset at the predetermined price during the agreed-upon expiration date.

Put options give the owner the *right* (not the obligation) to sell an underlying asset at an agreed-upon date at a predetermined price. *Sellers of put options receive an upfront premium payment from the put buyer* in exchange for the *obligation to buy* the underlying asset at the predetermined price during the agreed-upon expiration date.

## ASIAN / EUROPEAN / AMERICAN STYLE

Vanilla call and put options come in different styles, as outlined below:

**ASIAN OPTIONS**: Contracts that utilize the average price of the underlying asset over a predefined time when calculating the settlement value at expiration.

**EUROPEAN OPTIONS**: Contracts that can only be exercised at the option's expiration date and rely on an agreed-upon reference price for calculating the settlement value of the underlying asset at the time of the option contract expiration. European options are the most common style.

**AMERICAN OPTIONS**: Contracts that can be *exercised at any time*, meaning a call option owner could request the right to receive bitcoin delivery *anytime* from the contract's inception through the expiration date. American-style options are generally more expensive than other styles, which only allows the ability to exercise the option after expiration.

All else equal, if an underlying asset's price rises, THEN call options increase in value while put options decrease in value.

All else equal, if an underlying asset's price falls, THEN call options decrease in value while put options increase.

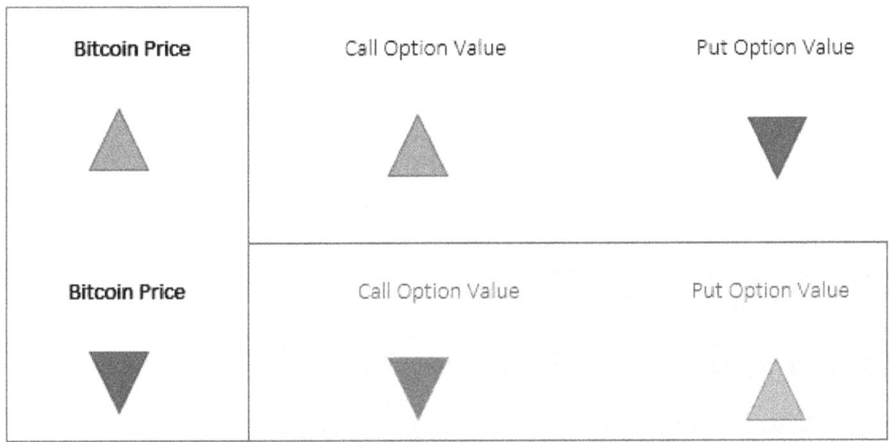


xxiv

## CONTRACT SPECIFICATIONS

The contract specifications below will be the fixed specifications for all options contracts discussed in this book.

**Contract Size**: 1 bitcoin

**Symbol**: BTC

**DURATION**: Daily, Weekly, Monthly, Quarterly, Yearly

**EXPIRATION DATE**: Daily (for daily options) or the last Friday of the duration.

**EXPIRATION TIME**: 8:00 UTC

**REFERENCE CURRENCY**: U.S.A. Dollars

**REFERENCE RATE**: Average of CME & CBOE Bitcoin Index

**STRIKE PRICE**: The price at which a transaction will occur if the buyer exercises the option.

**TYPE**: Call or Put

**STYLE**: European

**SETTLEMENT**: Physical bitcoin delivery or equivalent U.S.A. Dollar payment.

## READING THE MARKET

Below is a dataset for August 26th, 2022, BTC/USD options contracts that will be continuously referenced and analyzed. As options markets price from 0 DTE (days to expiration) to two years until expiration or even further, it is essential always to be aware of the expiration date of the analyzed options market. This data is a randomly selected BTC/USD monthly option market from 2022. On the top left, we see the BTC/USD reference rate of $23,093 at the time of the data snapshot. At the top center of the table is the expiration date (August 26th, 2022) of the listed options. To the right of that are the snapshot date and the time until expiration (August 2nd, 2022; **23 days, 9 hours to expiration**).

| BTC/USD 23,093 | | | | | 26-Aug, 2022 | | | | 2-Aug | 23D 9HR |
|---|---|---|---|---|---|---|---|---|---|---|
| CALL | | | | | | PUT | | | | |
| **IV BID** | **BID** | **ASK** | **IV ASK** | **DELTA** | **STRIKE** | **IV BID** | **BID** | **ASK** | **IV ASK** | **DELTA** |
| | | | | | 18000 | 84.1 | 254 | 277 | 86.3 | -0.10 |
| - | - | 5147 | 121 | 0.86 | 19000 | 79.7 | 369 | 393 | 81.5 | -0.15 |
| 74.1 | 3630 | 3741 | 81.1 | 0.80 | 20000 | 77.1 | 554 | 589 | 79.2 | -0.20 |
| 72.5 | 2890 | 3006 | 78.4 | 0.73 | 21000 | 74.6 | 809 | 843 | 76.4 | -0.28 |
| 72.4 | 2265 | 2346 | 76.1 | 0.64 | 22000 | 72.8 | 1156 | 1190 | 74.4 | -0.36 |
| **71.4** | **1722** | **1757** | **72.9** | **0.55** | **23000** | **71.7** | **1607** | **1641** | **73.2** | **-0.45** |
| 70.5 | 1271 | 1306 | 72 | 0.45 | 24000 | 70 | 2138 | 2184 | 72 | -0.55 |
| 69.1 | 901 | 947 | 71.3 | 0.36 | 25000 | - | - | 4647 | 150 | -0.64 |
| 68.8 | 635 | 670 | 70.5 | 0.28 | 26000 | - | - | 5318 | 150 | -0.72 |
| 68.5 | 439 | 462 | 69.9 | 0.21 | 27000 | - | - | 6024 | 150 | -0.79 |
| 68.6 | 300 | 323 | 70.2 | 0.16 | 28000 | 65.1 | 5132 | 5294 | 76.3 | -0.84 |
| 70.5 | 150 | 161 | 71.7 | 0.09 | 30000 | | | | | -0.91 |

xxv

Looking at the column labels above, the call option market is on the left side of the STRIKE column, while the put option market is on the right. Call options provide the right to own the asset on the expiration date at the strike price. Put options give the right to sell the asset on the expiration date at the strike price. BID & ASK columns represent the best U.S. dollar price for each contract as denoted by the strike price that buyers and sellers are willing to purchase for or sell at, respectively.

The Implied Volatility (IV) BID, and IV ASK columns represent the implied volatility for each option contract based on the best bid and ask prices from engaged market participants. DELTA signifies the rate of change for each option contract given a $1.00 move higher or lower for the BTC/USD reference rate, along with a shorthand reference to the *probability* each option currently has to be *in the money* with intrinsic value at expiration.

## THE FORWARD CURVE

Before we continue with options, we must introduce futures markets and their role in options trading. For a deep dive into futures, please refer to my book on futures contracts, which provides in-depth context, as this text will only scratch the topic's surface.

The forward curve in derivative markets, such as bitcoin options and futures contracts, is a forward market price structure formed through supply and demand forces driven by the ability to assume optionality or obligatory positions to either buy or sell bitcoin at a predefined price at some future date.

The spot market for bitcoin, from which derivative markets are built, is for bitcoin to be bought or sold and delivered on the same day. Since options and futures contracts are derivatives of the spot market, the delivery date of the asset occurs at the contract expiration date for European-style contracts. For example, the options market previously mentioned on August 2nd, 2022, is for BTC/USD with an expiration date of August 26th, 2022. Although not displayed here, those BTC/USD August 26th, 2022, options are accompanied by the BTC/USD September 30th, 2022, options market, BTC/USD October 28th, 2022, November 25th, 2022, options market, and so forth through time. In between, the major monthly expirations are the daily and weekly options markets.

Like options, futures contracts are derivatives of an underlying asset with related characteristics and critical differences. Futures markets allow market participants to ***lock in*** either *the sale price* or *the purchase price* of the underlying asset to be delivered or received at the contract's expiration date. There is no *optionality* component to holding a futures contract position, either long or short. Buying or selling futures contracts creates an *obligatory* position from the executed price to be settled at expiration. Futures contract positions can be offset and closed before expiry.

The price of futures contracts differs from expiration to expiration, typically trading at a premium to the spot market linearly from the nearest dated contract. Market participants can make a time series graph of futures market prices at different expiration dates, and the result is known as the ***forward curve***. The critical force behind the forward curve is the supply and demand for an underlying asset to be received or delivered at some point in the future and the *cost of carry* for that asset. Cost of carry is the analysis that must be done to understand the associated costs incurred when buying or selling a futures contract to either receive or deliver an underlying asset, like bitcoin, at some point in the future.

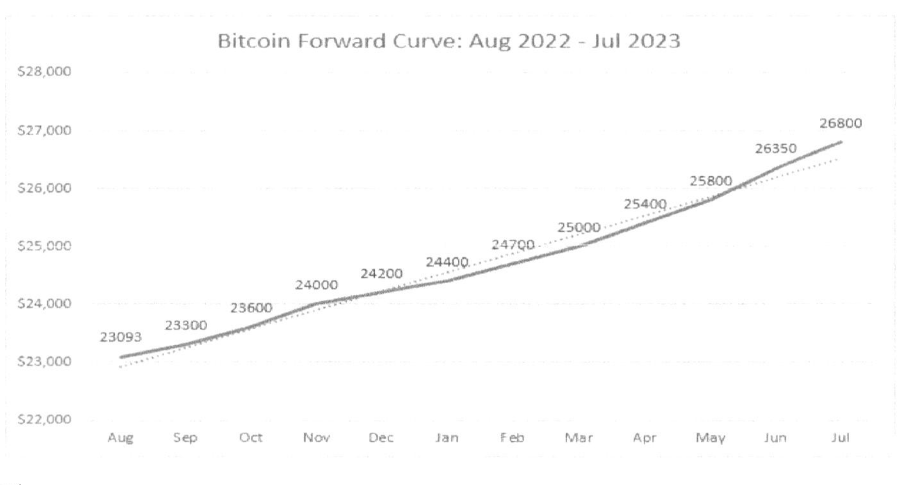


xxvi

Above we can see bitcoin prices to be delivered on a future date at a predetermined price with the spot market currently priced at $23,093 (Aug).

# COST OF CARRY

Treasury analysts often do cost-of-carry analysis for commodity producers and consumers, who then seek to engage with futures markets to *lock in* forward sale or purchase prices and delivery times that provide an acceptable rate of return for their core business. Managing costs and delivery dates can enhance operational efficiency, relieve bottlenecks, and provide clarity for balance sheet risk management. Treasury analysts managing balance sheet risk for gold miners are focused on net input costs to mine an ounce of gold relative to the market price for one ounce of gold. Therefore, understanding where the gold spot market is trading relative to the forward gold price to be delivered at some point in the future from the present moment is critical data for hedging and yield capture decisions.

Managers at oil refineries that input crude oil and output refined products like gasoline or diesel use futures market forward curves to determine profit margins. The price difference between input and output products can be locked in with futures contracts delivered or received at a predefined time. A crude oil refiner may seek to buy crude oil futures and simultaneously sell gasoline and diesel futures (known as the crack spread), allowing the refiner to lock in the purchase price of the primary input – crude oil, while also locking in the sale price of the refined products – gasoline and diesel.

For bitcoin miners, if the futures market trades at prices that allow for acceptable margins against their primary input cost – **electricity**, then bitcoin miners may sell forward to lock the price of bitcoin block rewards. The main input cost for bitcoin mining is electricity, which is usually locked in against a power purchase agreement that allows a bitcoin miner to know their exact cost per kilowatt-hour (kWh) of electricity.

Therefore, cost of carry analysis begins with net input costs calculations, including storage and delivery, for the production and distribution of the underlying asset, bitcoin. Since the forward curve technically sets the prevailing interest rate as an opportunity cost from the forward price of bitcoin relative to the spot price of bitcoin, cost of carry analysis also considers prevailing interest rates for the world reserve currency – U.S.A. treasury bond rates, which are considered risk-free from a counterparty perspective.

## FORWARD MARKET TERM STRUCTURE:

- **Contango**: When a futures market curve is positively sloping, the subsequent contracts are trading at a premium relative to the prior month's contract.
- **Backwardation**: When a futures market curve is negatively sloping, the subsequent contracts are trading at a discount relative to the prior month's contract.

## CONTANGO MARKETS

A bitcoin holder has a cost of carry analysis that differs from a crude oil refiner's cost of carry analysis. Bitcoin holders have storage and maintenance costs, albeit that is not the primary driver of the analysis. To fully grasp the bitcoin cost of carry, let's assume you are a holder and wish to generate additional income from the asset. By analyzing the futures market, were prices to be in Contango, market participants could buy spot and sell forward to capture a risk-free rate of return. This is known as basis trading. Although there are other avenues to do so, we will cover two main points here besides the forward sale strategy for which the *cost of carry* is analyzed.

- Sell call options against bitcoin holdings to receive option premium yield.
- Lend your bitcoin for interest to other market participants (short sellers, exchanges, brokers).
- Sell bitcoin forward in the futures market and lock in a sale price for delivery at expiration.

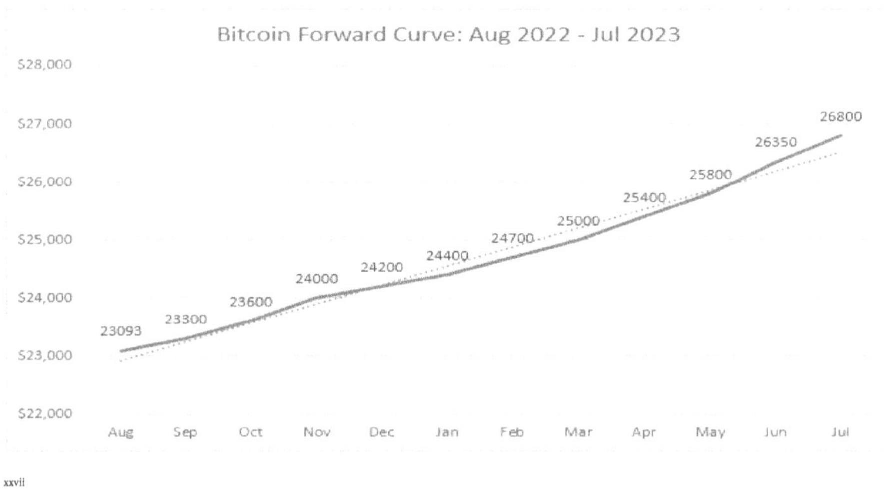


xxvii

# BASIS TRADE

When bitcoin futures contracts are trading at a premium to spot with the Contango market structure, that premium will erode through time as the expiration date of the contract approaches. The settlement price at contract expiration will be based on the spot market reference rate as notated in the contract specifications. This spread erosion is known as the ***basis trade*** or the **carry trade**. Spread erosion is demonstrated for the BTC/USD August 26th, 2022, futures contract price relative to the spot market price for bitcoin for the last 26 days of the contract's validity.

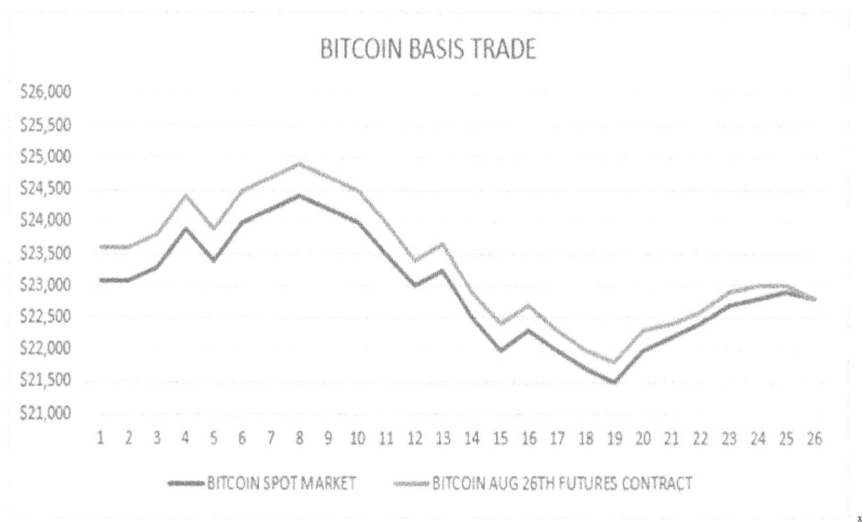


[xxviii]

In addition to bitcoin miners & HODLers seeking to lock in BTC/USD conversion rates at some point in the future, arbitragers will also seek to capture this spread through the purchase of bitcoin in the spot market, followed by a sale of the futures contract. This action generates a risk-free yield by locking the price difference between the purchased spot and forward sale prices. When bitcoin mining operations or asset managers decide to sell forward, lend bitcoin for yield through interest payments, or sell covered calls options, that yield becomes an opportunity cost input.

## FORWARD PRICE REFERENCE

When trading options, especially those with an expiration date greater than one month forward, it is critical to know where the futures market is pricing the corresponding futures contract to the option being analyzed and traded.

Due to a concept known as put-call parity, which will be discussed in the next chapter, the long position payoff of a futures contract can be replicated through the purchase of an ATM call option and the simultaneous sale of an ATM put option, with the same expiration date as the call (vice-versa to sell futures synthetically). Given this, **No significant arbitrage opportunities should exist between the futures contract price and the synthetic futures contract price that can be assembled with options contracts of corresponding expiration dates.**

The futures market price and the *at-the-money* options for a shared expiration date are synthetically linked. When analyzing options markets, the reference price is the corresponding futures contract price. For BTC/USD reference prices against the December 30th, 2022, bitcoin options market, the futures contract price for December 30th, 2022, would be referenced when pricing the options (as opposed to using the spot market price) to account for the forward curve impact on relative value between spot and forward prices.

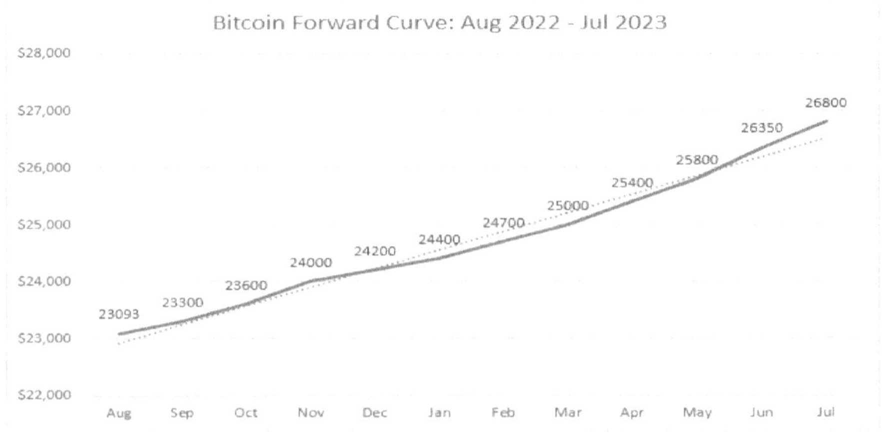

# INTRINSIC VS. EXTRINSIC VALUE

**Intrinsic value**: The value that a thing [option contract] has "in itself" or "For its own sake" or "as such" or "in its right[xxix]."

**Extrinsic value**: The value of something based on such things as appearance or what it could be sold for [in the future], which may not be its real [intrinsic] value:

A day trader's notion of value is determined by what another trader will pay a few moments later; it is extrinsic value[xxx].

- ***Intrinsic value*** in options: The difference between an options strike price and the underlying asset price.

**Options with intrinsic value are known as *IN THE MONEY*.**

- ***Extrinsic value*** in options: The difference between the market price of an option and its intrinsic value.

**Options with extrinsic value only are said to be *OUT OF THE MONEY*.**

| BTC/USD 23,093 | | | | | 26-Aug, 2022 | | | | 2-Aug | 23D 9HR |
|---|---|---|---|---|---|---|---|---|---|---|
| CALL | | | | | | PUT | | | | |
| IV BID | BID | ASK | IV ASK | DELTA | STRIKE | IV BID | BID | ASK | IV ASK | DELTA |
| | | | | | 18000 | 84.1 | 254 | 277 | 86.3 | -0.1 |
| - | 3759 | 5147 | 121 | 0.86 | 19000 | 79.7 | 369 | 393 | 81.5 | -0.15 |
| 74.1 | 3630 | 3741 | 81.1 | 0.8 | 20000 | 77.1 | 554 | 589 | 79.2 | -0.2 |
| 72.5 | 2890 | 3006 | 78.4 | 0.73 | 21000 | 74.6 | 809 | 843 | 76.4 | -0.28 |
| 72.4 | 2265 | 2346 | 76.1 | 0.64 | 22000 | 72.8 | 1156 | 1190 | 74.4 | -0.36 |
| **71.4** | **1722** | **1757** | **72.9** | **0.55** | **23000** | **71.7** | **1607** | **1641** | **73.2** | **-0.45** |
| 70.5 | 1271 | 1306 | 72 | 0.45 | 24000 | 70 | 2138 | 2184 | 72 | -0.55 |
| 69.1 | 901 | 947 | 71.3 | 0.36 | 25000 | - | - | 4647 | 150 | -0.64 |
| 68.8 | 635 | 670 | 70.5 | 0.28 | 26000 | - | - | 5318 | 150 | -0.72 |
| 68.5 | 439 | 462 | 69.9 | 0.21 | 27000 | - | - | 6024 | 150 | -0.79 |
| 68.6 | 300 | 323 | 70.2 | 0.16 | 28000 | 65.1 | 5132 | 5294 | 76.3 | -0.84 |
| 70.5 | 150 | 161 | 71.7 | 0.09 | 30000 | | | | | -0.91 |

[xxxi]

For example, looking at the BTC Aug 26th, 2022, $20,000 STRIKE Call option market, we see buyers willing to bid $3,630 for the right to buy bitcoin at $20,000 on Aug 26th, 2022. Given that the reference rate for bitcoin at the time of the snapshot was $23,093, the buyer(s) willing to pay $3,630 for the BTC AUG 26TH, 2022, $20,000 STRIKE CALL option is pricing in $3,093 of INTRINSIC value ($23,093 - $20,000) and $537 of EXTRINSIC value ($3,630 - $3,093).

Suppose the bitcoin reference rate equals $23,093. In that case, the $20,000 call option must be worth at least $3,093 (intrinsic value), or else there would be an arbitrage opportunity where a trader could buy the right to own bitcoin for less than $23,093 and then immediately sell bitcoin in the spot or futures market to capture a risk-free spread.

Any premium willing to be paid over intrinsic value is the option's extrinsic value, which can be considered a different variable that *could* make the option worth more than the current intrinsic value. Time is a significant component of extrinsic value, along with *implied volatility* (IV) of the underlying asset, which helps determine the market's expected range for the asset from the present moment through the expiration date. From the example above, where bitcoin is trading at $23,093, Aug 26, 2022, options expire in 23 days + 9 hours. Given the time to expiration, there is a chance that bitcoin can move higher in price before expiry, which would make call options more valuable. That potential is reflected in the *extrinsic* value of an option.

In addition to time and implied volatility to determine extrinsic valuations, traders will look at the historical price action of bitcoin and conclude that it is an asset that moves a lot, or in other words, there is a lot of *historical volatility* in the bitcoin price. These *known known* historical prices will influence the willingness of market participants to buy and sell options and the price at which transactions will occur. The price at which buyers are willing to bid, and sellers are willing to offer options will determine the value traders place on the *known unknown* of anticipated future price movement. *Implied volatility is that known unknown.* Using the variables of time and volatility, market participants can logically conclude why a trader may be willing to pay a higher price ($3,630) than just the option's intrinsic value ($3,093).

## AT THE MONEY

Options at the strike price that is closest to the BTC/USD reference rate are known as ***At-The-Money*** options. Since the reference rate for BTC/USD is $23,093, the 23000 strike CALL options are AT-THE-MONEY and have an intrinsic value of $93.00. We see buyers willing to pay $1,722 per option and sellers willing to accept $1,757.

The 23000 strike PUT options are the closest put options to being at-the-money but do not hold intrinsic value, only extrinsic value, given that the right to sell bitcoin at $23,000 has only time and implied volatility value since the current reference price is higher at $23,093. Despite only holding extrinsic value, buyers are willing to pay $1,607 for the option, and sellers are willing to accept $1,641 for the BTC/USD Aug 26$^{th}$, 2022, 23000 put option.

## INTRINSIC VALUE: IN THE MONEY OPTIONS

Call options with a strike price below the current reference rate are considered ***In The Money options,*** as the right to own an asset at a strike price below the current market price implies intrinsic value (reference rate – strike price).

On the contrary, put options that provide the right to sell an asset are considered *In The Money* when the strike price for the put option is above the current market reference rate, as the right to sell at a greater price than the reference rate implies intrinsic value (strike price – reference rate).

**IF** Call option intrinsic value = CURRENT REFERENCE RATE - STRIKE PRICE

And

**IF** Put option intrinsic value = STRIKE PRICE - CURRENT REFERENCE PRICE

**THEN** options contracts with positive intrinsic value are *In The Money.*

At the expiration date, only options that are ITM with intrinsic value will yield a profit to option buyers. The extrinsic value of an option is eroded through time from time decay itself. As the option approaches its expiration date, the utility of the option is no longer valid from that expiration date forward. This means that options that hold only extrinsic value and are OTM will become worthless on the expiration date unless the reference rate settles at a price that makes the option ITM with intrinsic value.

## RATIONAL EXERCISE

Bids and offers are the prices market participants are willing to buy or sell. Looking at the 23000-strike call options market, we see buyers willing to pay $1,722 per option for the right to own BTC/USD at $23,000 on Aug 26th, 2022, if rational to do so.

For the 23000-strike put options market, we see buyers willing to pay $1,607 for the right to sell BTC/USD at $23,000 on Aug 26th, 2022, if it is rational.

Were it to be rational for either the call buyer or the put buyer to exercise their right to buy or sell BTC/USD at 23,000 on Aug 26th, 2022, the option would be said to have expired IN THE MONEY.

Were it *irrational* for a call buyer or put buyer to exercise their optionality, it would be said that the options expired OUT OF THE MONEY with zero intrinsic value.

## Extrinsic Value

Out-of-the-money options are worthless at expiration. They derive present value primarily from time to expiry and the potential for the option to become ITM from future volatility experienced in the reference rate for the underlying asset's price.

| BTC/USD 23,093 | | | | | 26-Aug, 2022 | | | | 2-Aug | 23D 9HR |
|---|---|---|---|---|---|---|---|---|---|---|
| | CALL | | | | | | PUT | | | |
| IV BID | BID | ASK | IV ASK | DELTA | STRIKE | IV BID | BID | ASK | IV ASK | DELTA |
| | | | | | 18000 | 84.1 | 254 | 277 | 86.3 | -0.1 |
| - | 3759 | 5147 | 121 | 0.86 | 19000 | 79.7 | 369 | 393 | 81.5 | -0.15 |
| 74.1 | 3630 | 3741 | 81.1 | 0.8 | 20000 | 77.1 | 554 | 589 | 79.2 | -0.2 |
| 72.5 | 2890 | 3006 | 78.4 | 0.73 | 21000 | 74.6 | 809 | 843 | 76.4 | -0.28 |
| 72.4 | 2265 | 2346 | 76.1 | 0.64 | 22000 | 72.8 | 1156 | 1190 | 74.4 | -0.36 |
| **71.4** | **1722** | **1757** | **72.9** | **0.55** | **23000** | **71.7** | **1607** | **1641** | **73.2** | **-0.45** |
| 70.5 | 1271 | 1306 | 72 | 0.45 | 24000 | 70 | 2138 | 2184 | 72 | -0.55 |
| 69.1 | 901 | 947 | 71.3 | 0.36 | 25000 | - | - | 4647 | 150 | -0.64 |
| 68.8 | 635 | 670 | 70.5 | 0.28 | 26000 | - | - | 5318 | 150 | -0.72 |
| 68.5 | 439 | 462 | 69.9 | 0.21 | 27000 | - | - | 6024 | 150 | -0.79 |
| 68.6 | 300 | 323 | 70.2 | 0.16 | 28000 | 65.1 | 5132 | 5294 | 76.3 | -0.84 |
| 70.5 | 150 | 161 | 71.7 | 0.09 | 30000 | | | | | -0.91 |

xxxii

Looking at the BTC/USD Aug 26th, 2022, options market above, with a reference rate of $23,093, Call options with strike prices above 23,000 are considered out of the money but retain extrinsic value from time to expiration, and the potential for bitcoin price volatility before expiration. Put options with strike prices below $23,000 are considered out of the money but retain extrinsic value from time to expiry and the potential for bitcoin price volatility before expiration.

As time advances towards the option's expiration date, all option values will decrease relative to the option's time value decay (known as Theta), with a time value = 0 at the expiration time.

Rationally, an option contract must hold intrinsic value at the expiration time, or it will expire worthless. Given this, of the many listed options shown for the BTC/USD options market on August 26th, 2022, most option contracts will expire worthless and out of the money.

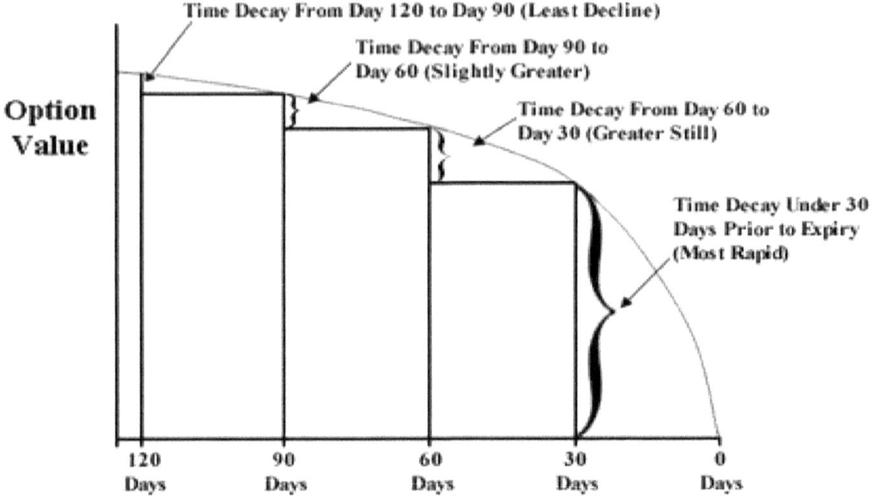


xxxiii

Further along in this text, the time value decay for an option contract will be examined. Known as Theta, time decay will be viewed in-depth as a variable used to identify potential opportunities that can generate yield & manage risk from selling options contracts.

When selling options, Theta becomes a positive value rather than a negative as when buying options. Sellers of options hold a beneficial position when the option sold expires worthless. Understanding the time decay of options contracts expands market participants' perspective on the effects of time, not simply related to reference rate price discovery but also to the value of optionality against the reference asset.

## BREAKEVEN PRICES

Sticking with the 23000 strike options with a reference rate of $23,093, we see call buyers willing to pay $1,722 for the right to buy bitcoin at $23,000 on Aug 26th, 2022. Given this information, we can add the option premium ($1,722) to the strike price ($23,000) to understand the price that bitcoin must be trading at for the call option buyer to break even on the position.

$1,722 (premium) + $23,000 (strike) = **$24,722**. This means bitcoin must be trading at $24,722 at the time of expiration for the call option owner to break even on the call option. To make money on the option, bitcoin would need to be trading above $24,722 at expiry. If the bitcoin settlement reference rate were between $23,001 and $24,722, the $23,000 strike call options would still hold intrinsic value for the buyer, but that value would be less than the net premium paid of $1,722 to buy the option. Were bitcoin reference prices to settle at $23,000 or below at expiration, the BTC/USD August 26th 23000 Call option would expire worthless with zero intrinsic value.

For the put option with a strike price of $23,000, we see buyers willing to pay $1,607 for the right to sell bitcoin at $23,000 on Aug 26th, 2022. Given this information, we can subtract the option premium ($1,607) from the strike price ($23,000) to understand the price that bitcoin must be trading at for the put option owner to breakeven, which is **$21,393**. For the put option holder to make a profit, bitcoin must be trading below $21,393 at the time of expiration.

The put option buyer in this example has paid $1,607 for the right to sell one bitcoin at $23,000 on August 26th, 2022. Given this premium cost, the price of bitcoin must be below $21,393 at the option's expiration date for the option's intrinsic value to be greater than the premium cost to buy the option. Were the bitcoin reference rate to settle between $22,999 and $21,393, the BTC/USD August 26th, 2022, 23000 put option would hold intrinsic value and be considered ITM. However, that intrinsic value would be less than the total premium cost to buy the option. **If** bitcoin settlement prices were $23,000 or higher, on August 26th, 2022, the 23000 put option would expire worthless, and the put option buyer would lose the entire premium paid for the option.

## PRICE DISCOVERY RANGE

With the information on breakeven prices for option buyers and the *moneyness* of an option being ITM, ATM, or OTM, we can realign our analysis to AT THE MONEY options, which are the 23000 strike calls and puts. Since the 23000-strike call option is bid for $1,722, call buyers need the market to move higher by 7.45% from the current reference price by the time of expiration to breakeven. This implies that call buyers anticipate that the market could increase by 7.45% or more in the next 23 days and 9 hours.

The 23000-strike put option is bid for $1,607; put buyers need the market to fall 6.95% by expiration to breakeven. This implies that put buyers anticipate that the market could move lower by 6.95%, or more, in the next 23 days and 9 hours.

| BTC/USD 23,093 | | | | | 26-Aug, 2022 | | | | 2-Aug | 23D 9HR |
|---|---|---|---|---|---|---|---|---|---|---|
| CALL | | | | | | PUT | | | | |
| **IV BID** | **BID** | **ASK** | **IV ASK** | **DELTA** | **STRIKE** | **IV BID** | **BID** | **ASK** | **IV ASK** | **DELTA** |
| 74.1 | 3630 | 3741 | 81.1 | 0.8 | 20000 | 77.1 | 554 | 589 | 79.2 | -0.2 |
| 72.5 | 2890 | 3006 | 78.4 | 0.73 | 21000 | 74.6 | 809 | 843 | 76.4 | -0.28 |
| 72.4 | 2265 | 2346 | 76.1 | 0.64 | 22000 | 72.8 | 1156 | 1190 | 74.4 | -0.36 |
| **71.4** | **1722** | **1757** | **72.9** | **0.55** | **23000** | **71.7** | **1607** | **1641** | **73.2** | **-0.45** |
| 70.5 | 1271 | 1306 | 72 | 0.45 | 24000 | 70 | 2138 | 2184 | 72 | -0.55 |
| 69.1 | 901 | 947 | 71.3 | 0.36 | 25000 | - | - | 4647 | 150 | -0.64 |
| 68.8 | 635 | 670 | 70.5 | 0.28 | 26000 | - | - | 5318 | 150 | -0.72 |
| 68.5 | 439 | 462 | 69.9 | 0.21 | 27000 | - | - | 6024 | 150 | -0.79 |

[xxxiv]

From the understanding of a market's anticipated price discovery range created by actual buyers and sellers willing to bid for ATM call & put options, market participants can use this practical perspective in parallel to assessing the mechanics of theoretical valuation. As we wrap up Chapter Two and prepare for an exploration into options pricing models, implied volatility, and options Greeks, Chapter Three provides a helpful yet theoretical approach to options markets. For practical application of options market analysis, it is critical to remain focused on prices stemming from actual buyers and sellers, not theoretical valuations alone.

CHAPTER

# MODELS, VOLATILITY & GREEKS

## BLACK-SCHOLES MODEL

The commencement of transactions at the Chicago Board of Options Exchange in 1973 coincided with the release of the paper, *The Pricing of Options and Corporate Liabilities,* published in the *Journal of Political Economy.* The paper was written by professors Fisher Black and Myron Scholes, who described an options pricing model based on variables that would be named in infamy as the *Black-Scholes* model. Through this relatively simple approach, a rush of academics and practitioners took an interest in this foundational options pricing model.

From Chicago to Hong Kong, options market participants spend years learning about and testing pricing theories against the Black-Scholes model and the historical outcomes of option market prices. To explore real yield and risk management opportunities via bitcoin options markets, a deep dive into option theory will be avoided. Theoretical assumptions will be examined from a high level only. The intent is to define a general understanding for practical application. The following image from *The Pricing of Options and Corporate Liabilities* expresses an equation to price a call option that does not need to incorporate the expected return of the underlying asset that the option is priced against.

Substituting from equation (12) into equation (9), and simplifying, we find:

$$w(x,t) = xN(d_1) - ce^{r(t-t^*)}N(d_2)$$

$$d_1 = \frac{\ln x/c + (r + \frac{1}{2}v^2)(t^* - t)}{v\sqrt{t^* - t}} \qquad (13)$$

$$d_2 = \frac{\ln x/c + (r - \frac{1}{2}v^2)(t^* - t)}{v\sqrt{t^* - t}}$$

In equation (13), $N(d)$ is the cumulative normal density function.

Note that the expected return on the stock does not appear in equation (13). The option value as a function of the stock price is independent of the expected return on the stock. The expected return on the option, however, will depend on the expected return on the stock. The faster the stock price rises, the faster the option price will rise through the functional relationship (13).

Note that the maturity $(t^* - t)$ appears in the formula only multiplied by the interest rate $r$ or the variance rate $v^2$. Thus, an increase in maturity has the same effect on the value of the option as an equal percentage increase in both $r$ and $v^2$.

Merton (1973) has shown that the option value as given by equation (13) increases continuously as any one of $t^*$, $r$, or $v^2$ increases. In each case, it approaches a maximum value equal to the stock price.

Providing further simplicity & clarity to the model:

$$C = SN(d_1) - Ke^{-rT}N(d_2)$$

$$d_1 = \frac{\ln(S/K)+(r+\sigma^2/2)T}{\sigma\sqrt{T}} \text{ and } d_2 = d_1 - \sigma\sqrt{T}$$

C = Theoretical call premium (European Style)

S = Current stock (asset) price

N = Cumulative standard normal distribution

K = Option strike price

r = Risk-free interest rate

T = Time remaining until option expiration

$\sigma$ = Volatility of the stock (asset)

## Greeks

Let $P$ refer to the equation for either a call or put option premium. Then the *greeks* are defined as:

Delta ($\Delta = \frac{\partial P}{\partial S}$): Where $S$ is the stock price.

Gamma ($\Gamma = \frac{\partial^2 P}{\partial S^2}$): Where $S$ is the stock price.

Theta ($\Theta = \frac{\partial P}{\partial t}$): Where $t$ is time.

Rho ($\rho = \frac{\partial P}{\partial r_f}$): Where $r_f$ is the risk-free rate.

Vega ($v = \frac{\partial P}{\partial \sigma}$) (Not Greek): Where $\sigma$ is volatility.

xxxv

Six critical variables go into the Black-Scholes model:

- The option's Strike Price is the price a transaction will execute if exercised.
- The current asset price (reference rate)
- Time to expiration
- Interest rates
- Annualized dividend yield
- Volatility of the underlying asset

There are seven *theoretical assumptions* that the model incorporates:

1. Short-term interest rates are known & constant through time.
2. Asset price discovery is random.
3. The asset yields no dividends during the options' time of expiration.

4. The option is European and can only be exercised at expiration.

5. No transaction costs to purchase or sell the asset or option.

6. There is sufficient liquidity to buy or borrow the asset at short-term interest.

7. The underlying asset can be sold short and settled with the buyer at a future date.

Notably, the assumptions above are simply that – assumptions. Models are not reality, despite their assistance in expressing logic.

## WINDY CITY PRICING MODELS

Although an inflection point in financial analytics, the Black-Scholes options pricing formula provides a model, there is no exact science to pricing options contracts. In practice, the price of an option is determined by willing buyers and sellers coming together through the motion of price discovery with bids and offers, where, eventually, a transaction will occur at a price agreed upon between the willing buyer and seller. Options pricing comes from science and art models to support pricing theory. They are designed to help solve for *known knowns, known unknowns,* and the acknowledgment of *unknown unknowns* relative to the future conversion rate for an underlying asset.

Aside from the approach, whether art, science, or hybrid-based, there are some baseline variables that most can agree upon as critical factors in pricing options. For example, it is essential to consider what dividend or yield payments (if any) will occur during the option's life and in what form of value the yield will be paid. Another would be *interest rates*; as an option buyer and option seller seek to agree on a price, the yield that could be earned on capital by investing in yield-bearing assets must be considered.

The storage and delivery costs of the underlying asset should also be thought through. Suppose the option seller has an obligation to deliver the underlying asset. Are there any costs associated with the storage and delivery of the asset if the buyer exercises the right to the option? For a call option buyer, what is the cost to take delivery if they exercise the right of an option and the ability to take physical possession of the underlying asset?

Options should also be priced considering any potential arbitrage opportunities. Options should not be priced to allow for a risk-free position to be assembled with a simultaneous transaction in the underlying asset's spot or futures market. *Time to expiration* and the underlying asset's historical price movement are critical factors in finding an option contract's fair market value.

## VOLATILITY

The variance of a dataset measures the range of data from the average value. The concept of *volatility* centers around understanding the range of a dataset over a predefined time (Y, Q, M, W, D, hour, minute). For BTC/USD, volatility measures the standard deviation of historical settlement prices. Standard deviation is the square root of the variance. How much do BTC/USD prices move up and down? Volatility provides a perspective.

Bitcoin volatility is the measure of how much the price of bitcoin varies over a defined period. It is calculated as the standard deviation of returns and represents the range within which bitcoin trades 68.2% of the time, which is a one-standard deviation range. *The Black-Scholes model uses volatility as the annualized standard deviation of lognormal price returns.* Lognormal returns account for a bitcoin floor price of $0.00. For futures markets where the reference rate can go negative, the normal distribution of price returns incorporates the potential for prices less than zero.

From an initial assessment of BTC/USD prices, many market participants are accustomed to looking at a chart and visualizing the prior price action. From there, the buyer determines where the asset's price could move to at some point in the future. Volatility takes the analysis further and applies mathematics to reinforce what our eyes see. Remember statistics class talked about probability distributions, the mean, median, variance, and standard deviation of a dataset, along with risk and reward analysis? I don't blame you if you slept through that lesson; it’s a notoriously dry topic in a classroom setting. However, these concepts come alive in the world of options. They can help manage risk and define real yield strategies with options and futures contracts.

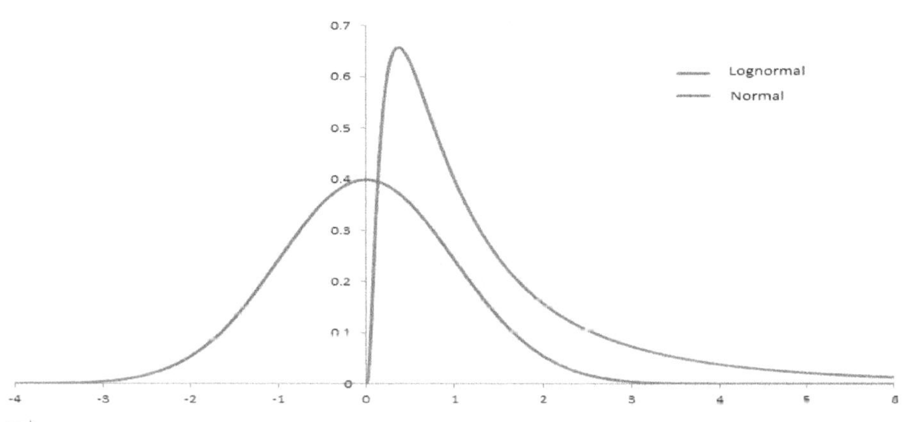


xxxvi

## HISTORICAL & REALIZED VOLATILITY

There are two main types of volatility: historical (realized) and implied (forward-looking):

***Historical Volatility (Realized)*** is the annualized standard deviation of lognormal returns from past asset price movements. More simply, historical volatility defines how much the asset moved up and down over the past *x* amount of time, assuming that prices cannot be below zero.

***Implied Volatility*** is a forward-looking and dynamic variable that estimates the anticipated future standard deviation of lognormal price distributions. **IV** is derived from willing buyers and sellers through actual bids and offers to buy or sell options. When there is no live market price, IV is an estimated variable.

To model the value of an option more accurately, IV is used as a primary input for the Black-Scholes formula despite the model's assumption that volatility is constant. If there is no transparent price for the option in question, then a trader could apply the historical volatility of the asset as the input for volatility or use the implied volatility from the option of a highly correlated asset.

Logically, *known knowns* can impact implied volatility. Known events in the future that could affect bitcoin prices, such as the Bitcoin protocol halving event or a legal ruling that may be anticipated from a national government or government regulators.

At the time of writing this book, there is speculation for a future ruling by the U.S.A. Securities and Exchange Commission (SEC) for a spot bitcoin ETF. This *known known* would be priced into implied volatility through supply and demand for options from active market participants.

## STANDARD DEVIATION

For historical volatility, the standard deviation of an asset's past price movements can be accurately calculated, which is helpful for market participants who seek to understand the asset's volatility over time and guide decisions on current volatility levels for the asset.

For market participants considering the purchase or sale of options in the present moment, implied volatility, or the expected annualized one-standard deviation price range for an asset, is of great importance given the insight into the market's expectation for future price movement ranges relative to the current reference price. As *growth rates* are *normally distributed*, probability distributions can help market participants understand the likelihood of where bitcoin prices could be at some point in the future.

For implied future bitcoin price range estimates from the current reference rate (0):

- One standard deviation represents the price range that bitcoin should trade roughly 68% of the time.
- Two standard deviations represent the price range bitcoin should trade roughly 95% of the time.
- Three Standard deviations represent the price range bitcoin should trade roughly 99% of the time.

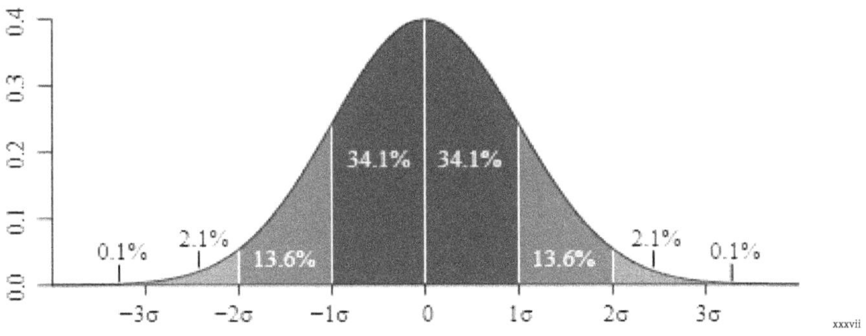

xxxvii

Bitcoin is an example of a highly volatile asset given the price range of BTC/USD encompasses wide price swings across multiple time horizons. In other words, bitcoin has a high standard deviation – the price range that bitcoin will trade in 68%, 95%, and 99% of the time is wider than that of a low standard deviation asset.

An example of a low volatility, low standard deviation asset would be McDonald's stock. It is a dividend-yielding asset with reasonably predictable cash flows, thus requiring more narrow price action to discover fair market value.

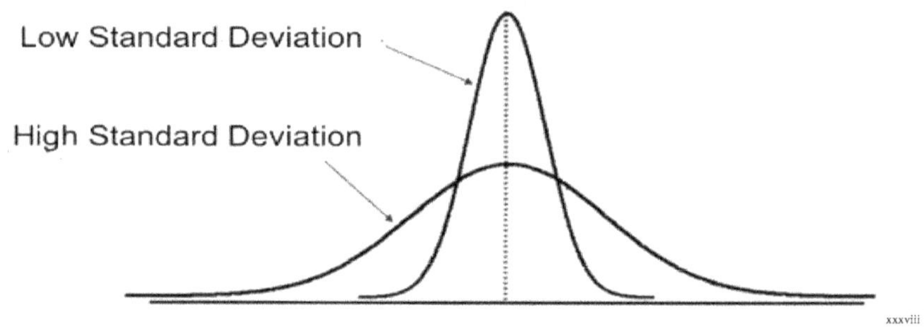


xxxviii

High Volatility (High Standard Deviation): Wide price swings and ranges.

Low Volatility (Low Standard Deviation): Narrow price swings and ranges.

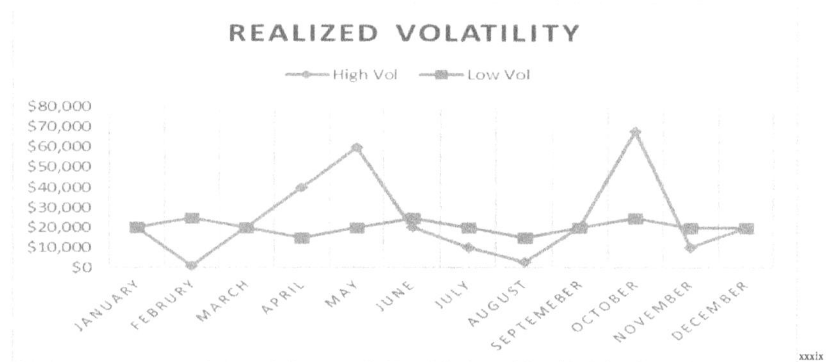


xxxix

# CALCULATING HISTORICAL VOLATILITY

Historical volatility (realized volatility) is the annualized standard deviation of lognormal price returns over a defined time. Realized volatility can be calculated for an underlying asset using historical price data.

The first step to calculating realized volatility is to compile time series data for the asset under analysis. For this example, we will examine 365 days of BTC/USD settlement price data from August 1, 2021, to August 1, 2022. Following the data compilation will be the calculation of daily normal returns for the series, then the lognormal returns.

BTC/USD

| Date | Settlement | Daily Returns | LN Returns |
|---|---|---|---|
| 8/1/2021 | $ 39,974.89 | #VALUE! | |
| 8/2/2021 | $ 39,201.95 | -1.9% | -2.0% |
| 8/3/2021 | $ 38,152.98 | -2.7% | -2.7% |
| 8/4/2021 | $ 39,747.50 | 4.2% | 4.1% |
| 8/5/2021 | $ 40,869.55 | 2.8% | 2.8% |
| 8/6/2021 | $ 42,816.50 | 4.8% | 4.7% |
| 8/7/2021 | $ 44,555.80 | 4.1% | 4.0% |
| 8/8/2021 | $ 43,798.12 | -1.7% | -1.7% |
| 8/9/2021 | $ 46,365.40 | 5.9% | 5.7% |
| 8/10/2021 | $ 45,585.03 | -1.7% | -1.7% |
| 8/11/2021 | $ 45,593.64 | 0.0% | 0.0% |

xl

Once lognormal returns are calculated for a BTC/USD settlement price dataset, realized volatility can be determined. Volatility is the standard deviation of lognormal historical price action. Excel can be used to compute the standard deviation of the lognormal returns. The example function is: =STDEV(D33:D398)

The returned value equals 3.6%, meaning from August 1st, 2021, to August 1st, 2022, the daily volatility for BTC/USD was 3.6%. In other words, one standard deviation of lognormal price distributions per day for BTC/USD during the past 365 days was 3.6% up or down from the prior day's settlement price. This means that roughly 68% of the time, bitcoin experienced a daily price range from August 1st, 2021, to August 1st, 2022, between -3.6% to +3.6% from the prior day's settlement price.

Furthering analysis, one could narrow the time series range from 365 days to the latest 28 days or a seven-day period, in which the goal would be to calculate realized volatility over the past x number of days rather than the prior year, as in the example above.

With pricing data for BTC/USD from August 1st, 2021 to August 1st, 2022, the daily volatility for BTC/USD was 3.6%. Multiplying this output by the square root of 365 days, the annualized historical volatility, or one standard deviation move for BTC/USD between August 1st, 2021, and August 1st, 2022, was 69%. In Excel, the SQRT function is helpful in this scenario. 0.036*SQRT(365) = **69%**

With annualized volatility at 69%, we can narrow the number to monthly, weekly, and daily volatility:

**Yearly volatility**: 69% * SQRT(1) = 69%

**Monthly volatility**: 69% * SQRT(1/12) = 20%

**Weekly volatility**: 69% * SQRT(1/52) = 13%

**Daily volatility** = 69% * SQRT(1/365) = 3.6%

These numbers can then be referenced against realized volatility for the past x number of days or against current implied volatility that the live market is pricing in to determine the relative value of volatility, which may present opportunities to participate in the purchase or sale of options contracts during the present moment.

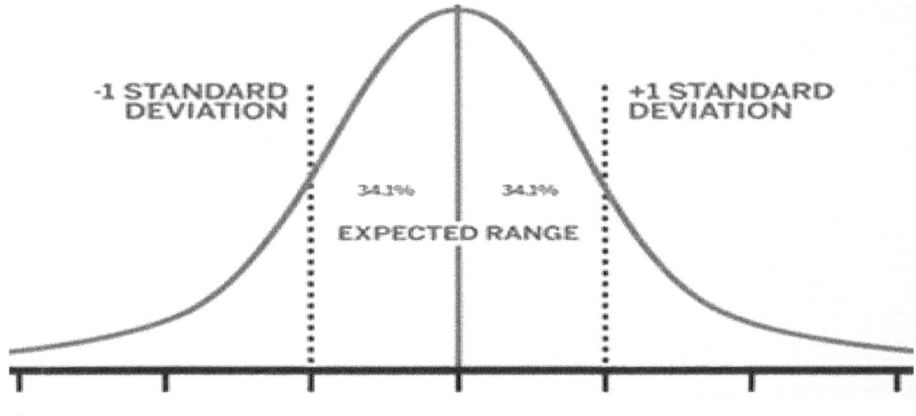


xli

# IMPLIED VOLATILITY

Implied volatility is the expected annualized standard deviation price range for an underlying asset at the present moment. Calculating implied volatility is to forecast the future price volatility of an underlying asset, which is a *guess,* just as weather forecasting is a guess. Options market participants and liquidity providers can actively trade implied volatility levels.

For *future* BTC/USD prices, the standard deviation states that:

- 1 standard deviation represents the price range that bitcoin should trade roughly 68% of the time.
- 2 standard deviations represent the price range bitcoin should trade roughly 95% of the time.
- 3 Standard deviations represent the price range bitcoin should trade roughly 99% of the time.

| BTC/USD 23,093 | | | | | 26-Aug, 2022 | | | | 2-Aug | 23D 9HR |
|---|---|---|---|---|---|---|---|---|---|---|
| | CALL | | | | | | | PUT | | |
| **IV BID** | **BID** | **ASK** | **IV ASK** | **DELTA** | **STRIKE** | **IV BID** | **BID** | **ASK** | **IV ASK** | **DELTA** |
| | | | | | 18000 | **84.1** | 254 | 277 | **86.3** | -0.1 |
| **-** | 3759 | 5147 | **121** | 0.86 | 19000 | **79.7** | 369 | 393 | **81.5** | -0.15 |
| **74.1** | 3630 | 3741 | **81.1** | 0.8 | 20000 | **77.1** | 554 | 589 | **79.2** | -0.2 |
| **72.5** | 2890 | 3006 | **78.4** | 0.73 | 21000 | **74.6** | 809 | 843 | **76.4** | -0.28 |
| **72.4** | 2265 | 2346 | **76.1** | 0.64 | 22000 | **72.8** | 1156 | 1190 | **74.4** | -0.36 |
| **71.4** | **1722** | **1757** | **72.9** | **0.55** | **23000** | **71.7** | **1607** | **1641** | **73.2** | **-0.45** |
| **70.5** | 1271 | 1306 | **72** | 0.45 | 24000 | **70** | 2138 | 2184 | **72** | -0.55 |
| **69.1** | 901 | 947 | **71.3** | 0.36 | 25000 | **-** | - | 4647 | **150** | -0.64 |
| **68.8** | 635 | 670 | **70.5** | 0.28 | 26000 | **-** | - | 5318 | **150** | -0.72 |
| **68.5** | 439 | 462 | **69.9** | 0.21 | 27000 | **-** | - | 6024 | **150** | -0.79 |
| **68.6** | 300 | 323 | **70.2** | 0.16 | 28000 | **65.1** | 5132 | 5294 | **76.3** | -0.84 |
| **70.5** | 150 | 161 | **71.7** | 0.09 | 30000 | | | | | -0.91 |

xlii

## EXPECTED RANGE

Referencing the BTC/USD AUG 26TH 2022, options market above, to the left of the BID column for both call and put options, and to the right of the ASK columns, are the implied volatility (IV BID, IV ASK) levels, respectively. Looking at the IV BID column, the 23,000-strike call option IV is currently 71.4%, based on the best bid price for that call option. This means the market participant(s) bidding $1,722 for the 23000-strike Aug 26th, 2022, call option is willing to buy the option at an implied volatility level of 71.4%, anticipating that the one-standard deviation range in the underlying asset (BTC/USD), up or down, will be 71.4% of the current market price of $23,093, or greater.

For call option buyers, the anticipation is predicated on the expectation for the price of BTC/USD to trend toward the upper bound of that standard deviation range. In contrast, put option buyers would take that position expecting the BTC/USD reference price to trend toward the lower bound of that range. Option sellers for both calls and puts would sell options with the expectation that realized volatility will be narrower than implied volatility. Call sellers can benefit from lower realized vol relative to implied vol at trade execution, with a tailwind if prices trend toward the lower bound. Put sellers benefit from lower realized vol relative to implied vol at the time of execution, with a tailwind if prices trend toward the range's upper bound.

Remember, in statistics class, a one-standard-deviation move is the price range an asset trades within 68% of the time. One standard deviation includes 68% of instances in a normal distribution, with 32% happening outside a one standard deviation range. With bitcoin priced at $23,093, an *implied* volatility of 71.4% means bitcoin has an *implied* one standard deviation price range between $6,605 and $39,581 during the following 365-day timeframe with a 16% chance of trading below $6,605 and a 16% chance of trading above $39,581, which would encompass two standard deviation and three standard deviation moves.

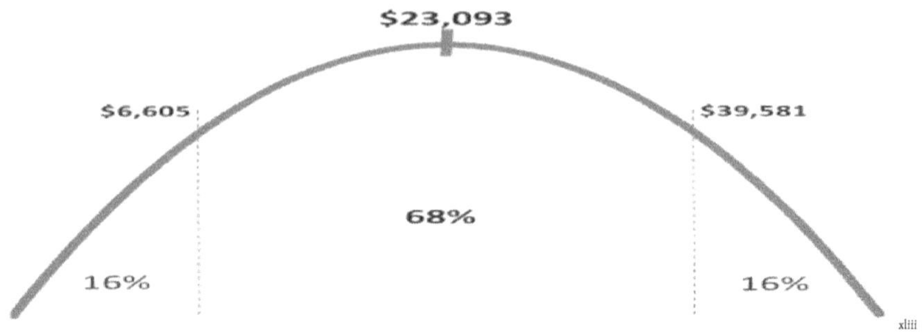


Bitcoin price * (IV/100) * SQRT (Days to Expiration/365)

$23,093 * (71.4/100) * SQRT (24/365)

$23,093 * .714 x 0.25 = $4,228

**EXPECTED RANGE UNTIL EXPIRATION** = $23,093 +/- $4,228 = **$18,865 - $27,321**

Note: An options calculator will provide a more accurate calculation of the expected range than the shorthand method above since, in practice, market prices experience a lognormal distribution, not a normal distribution. (spot bitcoin can only go to 0 but can appreciate to infinity).

This text does not focus on the study of implied volatility calculations, although the concept must be understood, even at a high level, when analyzing options markets. IV is a market expectation during a snapshot in time and moves up and down based on supply and demand for options, just as the underlying asset's price moves up and down based on supply and demand for that asset.

The table below examines different prices for a BTC/USD Aug 26th, 2022, 23,000-strike call option with a spot reference rate of $23,093 and roughly 24 days to expiration.

| BTC/USD | BTC AUG 26TH, 2022 23000 CALL | | |
|---|---|---|---|
| 23,093 | IV | PRICE | EXPECTED RANGE |
| | 100 | $2,411 | $17,172 - $29,014 |
| | **71.4** | **$1,722** | **$18,865 - $27,321** |
| | 50 | $1,239 | $20,133 - $26,053 |
| | 30 | $769 | $21,317 - $24,869 |
| | 10 | $302 | $22,501 - $23,685 |

[xliv]

From this visual, we can see what variable IV does to the price of an option, all else equal, and the expected range of the underlying asset. With this knowledge, we can now understand the concept of trading volatility and the speculation of an option's implied volatility being cheap or expensive.

## TRADING THE RANGE

A common options analysis compares implied volatility with realized (historical) volatility and determines whether the market is expensive or cheap on a relative basis compared to where volatility was previously realized. Again, there is no exact science in determining whether an option is cheap or expensive. However, relative analysis between current implied volatility and an underlying asset's historical or realized volatility can be a good baseline for those inclined to do their research.

The higher an option's implied volatility, the more the range of prices for a one-standard-deviation move expands. In comparison, a lower implied volatility level tightens the range of prices for a one-standard-deviation move. When we change the IV variable, the option's price and the expected range of the underlying asset from trade inception until expiry dramatically differs from an implied volatility of 10% to 100%.

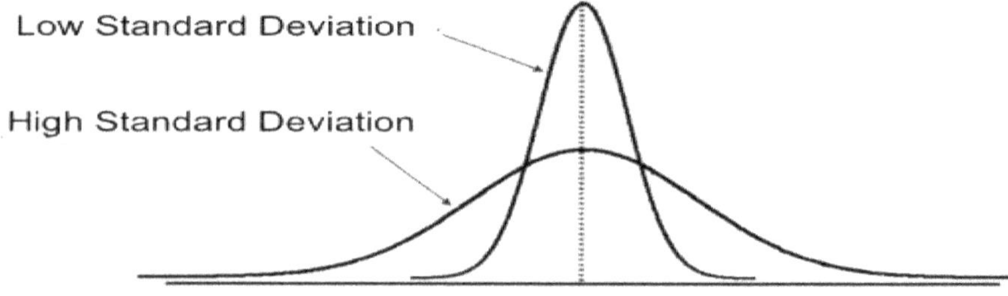


[xlv]

| BTC/USD | BTC AUG 26TH, 2022 23000 CALL | | |
|---|---|---|---|
| 23,093 | IV | PRICE | EXPECTED RANGE |
| | 100 | $2,411 | $17,172 - $29,014 |
| | **71.4** | **$1,722** | **$18,865 - $27,321** |
| | 50 | $1,239 | $20,133 - $26,053 |
| | 30 | $769 | $21,317 - $24,869 |
| | 10 | $302 | $22,501 - $23,685 |

xlvi

As shown in the table above, for the August 26, 2022, 23000 strike BTC/USD call option market displays a buyer is willing to pay $1,722 for the call option. The implied volatility is 71.4%, meaning the underlying asset's expected annualized price range is 71.4% of the bitcoin reference rate of $23,093 ($6,605 - $39,581). The one standard deviation price range for the bitcoin reference rate with 24 days until option expiration on August 26th, 2022, is +/- $4,228 ($18,865 - $27,321) from the current price of $23,093.

## THE GREEKS

The concept of the "Greeks" was first introduced to the financial world with the infamous paper from Fischer Black and Myron Scholes that we continue to reference. The Black-Scholes model applies mathematics to calculate the value of an option and the sensitivity of an option's price relative to changes in the underlying assets' price, time, volatility, and interest rates.

The *Greek* symbols (although some are not Greek) were probably selected from other mathematical and scientific notations that use Greek symbols, along with a nod to our friend Thales of Miletus, credited with conceiving and executing the first options contract in recorded history. The Greeks are used to define partial derivatives, meaning the derivative of a variable (price, time, volatility, interest rates) used in a mathematical function like the Black-Scholes model.

Below are the Greeks most often used in measuring an option contract's price sensitivity to external factors:

- **DELTA:** The rate of change to an options price relative to changes in the reference rate for bitcoin.
- **GAMMA**: The rate of change for Delta is relative to changes in the bitcoin reference rate.
- **THETA**: The rate of change of an option's price relative to time until expiration, also known as *time decay.*
- **RHO**: The rate of change in an option's price relative to changes in interest rates.
- **VEGA**: The rate of change of an option price relative to changes in implied volatility. Note that in the Black-Scholes model, an option's volatility is assumed to be constant, so in practice, Vega represents the change in an option's price for each 1% move in implied volatility.

Refraining from an all-encompassing expedition into functions and derivatives, the practical application of the Greeks will be explored through portfolio risk management and real yield strategies that can be obtained from bitcoin options markets.

## OPTION PRICING CALCULATORS

With the advent of modern computing technology, a wide variety of pricing calculators have been built to solve for Greek values. In addition, the expected range for one standard deviation moves from the trade date until the expiration date of an option contract is often included. You can search the internet and choose the calculator that suits you best.

Option pricing models depend on *known knowns* and *known unknowns*. For known inputs, such as the current price of bitcoin, the strike price, the trade date, the type of option, dividend yield, interest rates, and the option's expiration date, those inputs are straightforward.

Market participants must make an educated guess for the implied volatility variable for inputs that are known unknowns or things that we know that we don't know, such as the future volatility of the underlying asset.

When an option contract has live bids and offers from buyers and sellers, thus setting a bid price, mid-price, and ask price, you can back out the implied volatility in which the market is currently pricing. For this reason, most options venues display the implied volatility for bid, mid, and ask prices where there are live bids and offers.

| BTC/USD AUGUST 26th, 2022, 23000 CALL OPTION ANALYSIS | | | |
|---|---|---|---|
| | | | |
| BITCOIN PRICE | $23,093 | DIVIDEND YIELD | 0% |
| MODEL | BLACK-SCHOLES-MERTON | VOLATILITY | 71% |
| STRIKE PRICE | $23,000 | STYLE | EUROPEAN |
| TRADE DATE | Tuesday, August 2, 2022 | INTEREST RATE | 3% |
| TYPE | CALL | EXPIRATION DATE | 8/26/2022 |

xlvii

It's worth noting that an options calculator generates the price of an option based on several input factors. These include the reference price of bitcoin, time to expiration, and implied volatility, which can significantly impact the output values. It's essential to understand that minor changes to these inputs can cause a considerable shift in the option's price, while changes in the interest rate have a comparatively minor impact.

Additionally, thanks to put-call parity, it's possible to extrapolate the put option value when solving for a call option value. By comprehending these intricate details, one can make more informed decisions when trading options.

| AUGUST 2nd, 2022, 23000 OPTION GREEKS | | |
|---|---|---|
| 24 DAYS TO EXPIRATION | CALL | PUT |
| PRICE | $ 1,740 | $ 1,601 |
| DELTA | 0.55 | -0.45 |
| GAMMA | 0.000094 | 0.000094 |
| THETA | -35.57 | -33.68 |
| VEGA | 23.44 | 23.44 |
| RHO | 7.19 | 7.89 |
| OMEGA | 7.29 | -6.49 |

xlviii

Further, many calculators will return the one standard deviation expected range for the underlying asset from the trade inception date through the expiration date based on the assumption of annualized implied volatility. The calculator's expected range should be more accurate compared to shorthand calculations.

| ONE STANDARD DEVIATION EXPECTED RANGE \| VOLATILITY = 71% | | |
|---|---|---|
| BITCOIN PRICE | CALL OPTION PRICE | TIME PREMIUM |
| $ 18,888 | $ 274 | $ 274 |
| $ 19,729 | $ 439 | $ 439 |
| $ 20,570 | $ 663 | $ 663 |
| $ 21,411 | $ 953 | $ 953 |
| $ 22,252 | $ 1,311 | $ 1,311 |
| **$ 23,093** | **$ 1,740** | **$ 1,647** |
| $ 23,933 | $ 2,234 | $ 1,300 |
| $ 24,774 | $ 2,790 | $ 1,015 |
| $ 25,615 | $ 3,400 | $ 785 |
| $ 26,456 | $ 4,059 | $ 602 |
| $ 27,297 | $ 4,757 | $ 460 |

# VOLATILITY IMPACT

From the options calculator data above for the BTC/USD August 26, 23000-strike call option with 24 days until expiration and an implied volatility of 71%, we can see the one standard deviation expected range for the underlying asset from trade inception date through expiration to be **$18,888 - $27,297** when the reference price is **$23,093**.

All else equal, the data below shows the changes to the expected range of the underlying asset if the implied volatility is adjusted to 100% and 50%, respectively. Also noted are the modifications to an options price and the extrinsic (time premium) value of the option with volatility estimated or priced in by the live market at 100% and 50%, respectively.

| ONE STANDARD DEVIATION<br>EXPECTED RANGE \| VOLATILITY = 100% | | |
|---|---|---|
| BITCOIN PRICE | CALL OPTION PRICE | TIME PREMIUM |
| $ 17,171 | $ 325 | $ 326 |
| $ 18,355 | $ 553 | $ 553 |
| $ 19,540 | $ 870 | $ 871 |
| $ 20,724 | $ 1,286 | $ 1,286 |
| $ 21,908 | $ 1,802 | $ 1,802 |
| **$ 23,093** | **$ 2,418** | **$ 2,325** |
| $ 24,277 | $ 3,127 | $ 1,850 |
| $ 25,461 | $ 3,920 | $ 1,459 |
| $ 26,645 | $ 4,788 | $ 1,142 |
| $ 27,830 | $ 5,719 | $ 889 |
| **$ 29,014** | $ 6,704 | $ 690 |

| ONE STANDARD DEVIATION<br>EXPECTED RANGE \| VOLATILITY = 50% | | |
|---|---|---|
| BITCOIN PRICE | CALL OPTION PRICE | TIME PREMIUM |
| $ 20,132 | $ 219 | $ 219 |
| $ 20,724 | $ 337 | $ 337 |
| $ 21,316 | $ 495 | $ 495 |
| $ 21,908 | $ 698 | $ 698 |
| $ 22,500 | $ 948 | $ 948 |
| **$ 23,093** | **$ 1,247** | **$ 1,154** |
| $ 23,685 | $ 1,592 | $ 907 |
| $ 24,277 | $ 1,981 | $ 704 |
| $ 24,869 | $ 2,410 | $ 540 |
| $ 25,461 | $ 2,873 | $ 411 |
| **$ 26,053** | $ 3,366 | $ 312 |

# MULTIVARIABLE VALUATION

With the spot market price of bitcoin and futures contracts that settle against a bitcoin spot reference rate, buyers' and sellers' supply/demand function in the marketplace drives the price higher or lower. It sets the value for the underlying asset spot reference price or the futures contract reference price.

As unveiled through the Black-Scholes model and the corresponding options pricing calculators that leverage the model, options prices are multivariable. This means that the underlying asset's reference price alone does not set the value for the option. Instead, the reference price is combined with variables for time to expiration, volatility, and interest rates to determine the net value of an option. The reference price can remain constant, and the price of an option is still vulnerable to moving higher or lower, depending on shifts in the expected volatility, the prevailing interest rate, and the known known of time decay.

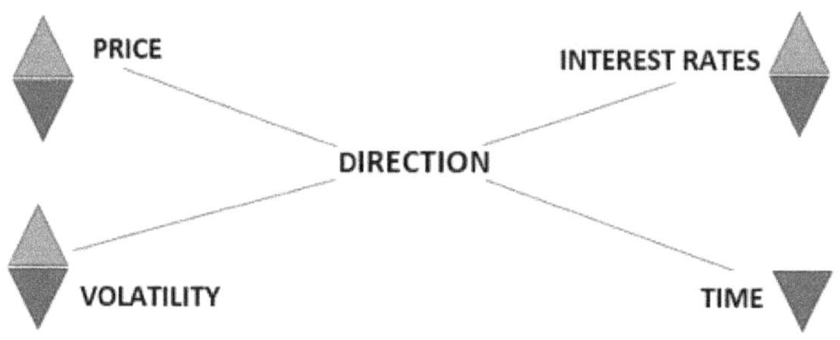

## DELTA

Delta is the rate of change for an options price relative to changes in the reference price of BTC/USD. For every $1.00 move in bitcoin prices, the value of an option contract will move higher or lower by the option's delta. The value of delta ranges between 0-1 for a long call option (or short put) and from -1-0 for a long-put option (or short call).

Since Delta is measured relative to the underlying asset's price, for every $1.00 move in the reference price, the option value will change by the Delta, all else equal. For example, IF the Aug 26th, 2022, BTC/USD 23000 CALL option has a Delta of 0.55, and the reference rate for BTC/USD = $23,093, THEN were the price of bitcoin to move higher by $1.00 from $23,093 to $23,094, the price of the call option would increase from $1,722 to $1,722.55, all else equal. **If** the bitcoin price were to move lower by $1.00, **then** the call option price would decrease by the Delta from $1,722 to $1,721.45.

For the put option with the same strike price and expiration date, the Delta is -0.45. **If** the price of BTC/USD rose from $23,093 to $23,094, the put option price would decrease by $0.45 from $1,607 to $1,606.55. **If** the price of BTC/USD fell from $23,093 to $23,092, **then** the value of the put option would increase by $0.45 from $1,607 to $1,607.45.

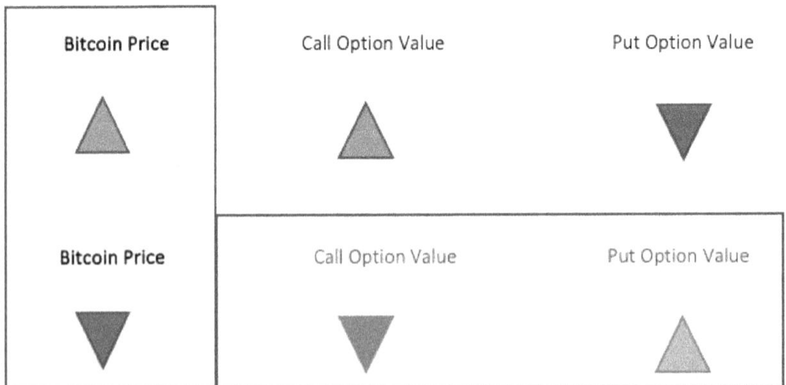

# DELTA AS PROBABILITY

When discussing options being *in the money*, *out of the money*, or *at the money*, we can think of the question being associated with the *moneyness* of an option. *Moneyness* is the relative positioning between the reference rate and the call and put options that trade against that price.

To refresh, call options with a strike price below the reference rate and put options with strike prices above the reference rate are considered in the money, as they hold intrinsic value. Call options with a strike price above the reference rate and put options with a strike price below the reference rate are considered *out of the money*, as they only hold time value (extrinsic value). Call and put options with a strike price equal to, or very close to, the reference rate are said to be *at the money*.

Moneyness seeks to quantify how far an option is either ITM, OTM, or ATM by solving for the implied probability that an option will expire ITM, expressed as a percentage. **Although not exact, Delta is a close approximation of ITM probability at expiration**.

In theory, markets are random – meaning there is a 50/50 chance that the price of an asset will be higher or lower tomorrow or at any time in the future relative to the asset's price today. Given this, Delta as the percent chance of an option being in the money or out of the money at the time of expiration should be 0.50 for call options that are ATM and -0.50 for put options that are at the money. ATM strikes are nearly or exactly the reference rate, which, according to the random markets theory, has a 50/50 chance of moving higher or lower. Delta for ITM options trend toward 1 for calls and -1 for puts, and OTM options should have a delta that is less than 0.50 for calls and greater than -0.50 for puts, trending toward 0.

For example, the Aug 26th, 2022, BTC/USD 23000 CALL option has a Delta of 0.55. This means that for every $1.00 move in BTC/USD, the call option price will move by $0.55, all else equal. **The value of Delta can also be expressed as the percent likelihood of an option being ITM at the time of expiration**. Therefore, the Aug 26th, 2022, BTC/USD 23000 CALL option with the spot reference rate at $23,093 and a Delta of 0.55 has a 55% chance of expiring ITM. The corresponding put option has a 45% chance of expiring ITM, as same strike, same expiration call, and put options have absolute value Deltas that sum to 1.

As the price of BTC/USD moves up and down through time, the option's Delta changes relative to the option's Gamma. The probability that an option will expire ITM changes as the price of the underlying asset changes, with the option either becoming further ITM, OTM, or remaining near ATM.

# PUT-CALL PARITY

Before we continue with the Greeks, it's essential to finish our understanding of Delta as the rate of change in an options price relative to changes in the reference rate. Delta is also a shorthand reference to the percent chance that the option will expire ITM. By examining Delta between call and put options with the same strike price and expiration date, that relationship leads to the unveiling of put-call parity.

| BTC/USD 23,093 | | | | | 26-Aug, 2022 | | | | 2-Aug | 23D 9HR |
|---|---|---|---|---|---|---|---|---|---|---|
| CALL | | | | | | PUT | | | | |
| IV BID | BID | ASK | IV ASK | DELTA | STRIKE | IV BID | BID | ASK | IV ASK | DELTA |
| 72.4 | 2265 | 2346 | 76.1 | **0.64** | 22000 | 72.8 | 1156 | 1190 | 74.4 | **-0.36** |
| **71.4** | **1722** | **1757** | **72.9** | **0.55** | **23000** | **71.7** | **1607** | **1641** | **73.2** | **-0.45** |
| 70.5 | 1271 | 1306 | 72 | **0.45** | 24000 | 70 | 2138 | 2184 | 72 | **-0.55** |

Notice the 23000-strike, ATM call, and put options. The call option has a Delta of 0.55, while the put option has a Delta of -0.45. Now look at the 22000-strike call and put Deltas. Notice the call Delta is 0.64 while the put Delta equals -0.36. Further, the 24000-strike call and put options have Deltas of 0.45 and -0.55, respectively. Intuitively, we can take the absolute value of call Delta added to the absolute value of the put Delta, where in each instance, the sum is equal to 1. This is the first step to understanding put/call parity.

**If** a call option is the right to buy the underlying asset at the strike price, and a put option is the right to sell the asset at the strike price, **then** being long a call and short a put (+call-put) replicates the payoff of owning the forward price of an underlying asset. Therefore, buying ATM calls and selling ATM puts is roughly equivalent to the forward price of the asset, which has a Delta = 1. To replicate the payoff of a short forward position, traders can do the inverse, buy puts, and sell calls with the same strike (-call +put).

Since a long call option has a positive Delta, for every $1.00 move higher in the underlying asset, the value of the call increases by the Delta; all else equal.

Short puts also have a positive Delta (since long puts have a negative Delta). For every $1.00 rise in the underlying asset, the put option decreases by the Delta value. Decreases in the option price are favorable for the option seller. Therefore, short puts have a positive Delta, while long puts have a negative Delta. Long calls have positive Delta, whereas short calls have negative Delta.

If a call option is purchased, and the equivalent amount of the underlying asset is sold relative to the Delta of a call, then a 0.55 Delta call is neutralized by selling 0.55 * 1 bitcoin. That portfolio would have the same payoff as selling the same strike and expiration put option while selling the underlying asset relative to the put option Delta. Selling a 0.45 Delta put is neutralized through selling 0.45 * 1 bitcoin. The same can also be said for a short call option (Delta neutralized through the purchase of the underlying asset) and a long put, which is Delta neutralized through the purchase of the underlying asset with respect to the options Delta.

**Delta Neutral Portfolio**

The short-term price risk of an options position can be offset by the simultaneous purchase or sale of the underlying asset (bitcoin) in relation to the call options Delta. A 0.50 delta option requires 0.50 of the underlying to be bought or sold to neutralize the short-term price risk of the option, depending on the option type, either call or put, and the direction of the position, either long or short.

- **Long call** option can be neutralized by simultaneously selling the underlying asset.
- **Short call** option can be neutralized through the simultaneous **purchase** of the underlying asset.
- **Long-put** option can be neutralized by simultaneously purchasing the underlying asset.

- **Short put** option can be neutralized by simultaneously selling the underlying asset.

## GAMMA

Gamma is the second derivative of Delta. Effectively, Gamma is the delta of Delta. Delta is the rate of change of an options price relative to changes in the reference rate. As the delta of Delta, Gamma is the rate of change of Delta. When Delta is increasing, Gamma is the acceleration value of that increase. When Delta is decreasing, Gamma is the deceleration value of that decrease.

Delta is dynamic and changes with movements in the underlying asset. Gamma, as the sensitivity to the rate of change of Delta, is the measurement that shows how much Delta will change, either up or down, in correlation to the movement of the underlying asset.

For example, Suppose BTC/USD is currently priced at $23,093, and the BTC/USD Aug 26th, 2022, 23000 Call option has a price of $1,722 and a delta of 0.55. In that case, it is understood the price of the option will increase or decrease by $0.55 for every $1.00 move, either up or down in the underlying asset respectively. Now, let's add Gamma. It's the same scenario and option. However, we will incorporate the Gamma value for the option, which, for example, will be 0.01. **If** BTC/USD were to move higher by $1.00 from $23,093 to $23,094, **then** the call option value would increase by $0.55, and Delta would then change by the value of Gamma from 0.55 to 0.56. With BTC/USD now at $23,094, the new Delta for the call option equals 0.56, and subsequent moves in the underlying asset will adjust the price of the call option by the new Delta or $0.56 for every $1.00 move in the underlying asset.

An option with positive Delta, such as a long call, increases in value when the reference rate rises. At the same time, an option with a negative Delta, such as a long put, decreases when the reference price increases. Since option values are estimated to rise and fall by the value of their Delta relative to a $1.00 move up or down in the underlying asset, and Delta is dynamic, Gamma can be considered as the rate of change in Delta during reference rate movements. Delta rises or falls as the likelihood of the option expiring ITM moves higher or lower. Gamma is the acceleration or deceleration of Delta and is dynamic, adjusting to reference rate movements just as Delta is dynamic and adjusts by Gamma when there are movements in the reference rate.

# VEGA

Vega measures an option contract's price sensitivity to changes in implied volatility. Vega is the value an option will gain or lose as implied volatility moves higher or lower by 1%.

Given that implied volatility sets the expected range for the underlying asset, all options prices rise to the value of their Vega if implied volatility rises. As the expected range widens, the probability of options expiring in the money increases. Conversely, falling implied volatility causes all options to lose value relative to the value of Vega, given that lower implied volatility implies a tighter expected price range for the underlying asset. The tightening of the expected range depreciates the probability that options expire in the money.

Consider an option portfolio of a long ATM call option and a long ATM put option, otherwise known as a *straddle* position. By owning the optionality to either buy or sell at the same strike price, each option becomes more valuable as implied volatility increases. This increased probability is reflected in Vega, which will increase the price of each option for every 1% change higher in implied volatility.

When engaged in a *strangle* position, where a portfolio of an OTM call and OTM put option is constructed, this structure has a high degree of sensitivity to changes in implied volatility. Given the portfolio of OTM call and put options, **if** implied volatility were to increase, **then** the likelihood of options expiring in the money would increase. Thus, Vega represents the price impact each option experiences from a 1% change in implied volatility. Were implied volatility to fall, the likelihood of options expiring in the money declines, and the value of an option decreases by the value of Vega for every 1% move lower in implied volatility.

# THETA

Longer-dated options contracts provide more opportunity for the option to expire ITM with intrinsic value than shorter-dated options contracts. As time progresses, options lose value each day relative to the Theta value of an option. Given this daily time decay, Theta is represented as a negative number.

| AUGUST 2nd, 2022, 23000 OPTION GREEKS | | |
|---|---|---|
| | | |
| 24 DAYS TO EXPIRATION | CALL | PUT |
| | | |
| PRICE | $ 1,740 | $ 1,601 |
| DELTA | 0.55 | -0.45 |
| GAMMA | 0.000094 | 0.000094 |
| THETA | -35.57 | -33.68 |
| VEGA | 23.44 | 23.44 |
| RHO | 7.19 | 7.89 |
| OMEGA | 7.29 | -6.49 |

From our options price calculator above, for the August 26, 2022, 23000 CALL option, we can see that Theta for the option on August 2$^{nd}$, 2022 (the initial trade date) was priced at -35.57. This means the option's value will lose $35.57 for each passing day due to the expiration date getting closer by one day. The slope of decay for theta accelerates as the option gets closer to its expiration date. The value of Theta equals zero at expiry.

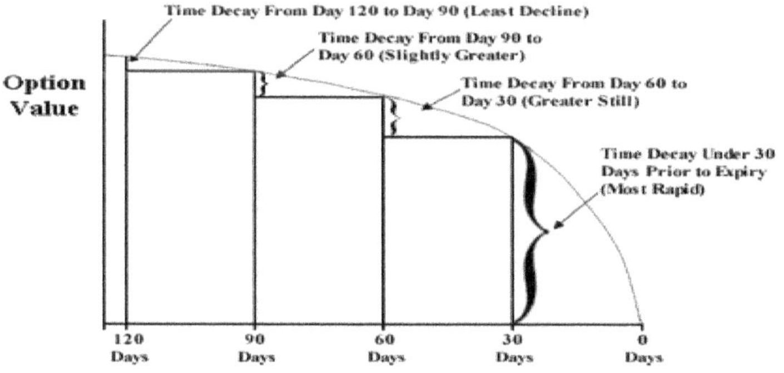


xlix

## RHO

The Greek variable for Rho represents the value an option will change, either higher or lower, relative to changes in interest rates.

## OMEGA

Although a lesser-known Greek variable, Omega rounds out our Greek analysis. Omega measures the elasticity of an option or the percent change in the value of an option per percent change in the reference rate for the underlying asset. Omega can help measure risk, as the higher the Omega value, the more movement an options price will experience relative to changes in the reference rate of the underlying asset.

| BTC/USD AUGUST 26th, 2022, 23000 CALL OPTION ANALYSIS | | | |
|---|---|---|---|
| BITCOIN PRICE | $23,093 | DIVIDEND YIELD | 0% |
| MODEL | BLACK-SCHOLES-MERTON | VOLATILITY | 71% |
| STRIKE PRICE | $23,000 | STYLE | EUROPEAN |
| TRADE DATE | Tuesday, August 2, 2022 | INTEREST RATE | 3% |
| TYPE | CALL | EXPIRATION DATE | 8/26/2022 |

| AUGUST 2nd, 2022, 23000 OPTION GREEKS | | |
|---|---|---|
| 24 DAYS TO EXPIRATION | CALL | PUT |
| PRICE | $ 1,740 | $ 1,601 |
| DELTA | 0.55 | -0.45 |
| GAMMA | 0.000094 | 0.000094 |
| THETA | -35.57 | -33.68 |
| VEGA | 23.44 | 23.44 |
| RHO | 7.19 | 7.89 |
| OMEGA | 7.29 | -6.49 |

## PRICING THE OPTION

Below, we refocus our attention on the original options market that we continuously reference, now armed with knowledge of the Black-Scholes options pricing model, the Greeks, and implied volatility.

| BTC/USD 23,093 | | | | | 26-Aug, 2022 | | | | 2-Aug | 23D 9HR |
|---|---|---|---|---|---|---|---|---|---|---|
| CALL | | | | | | PUT | | | | |
| IV BID | BID | ASK | IV ASK | DELTA | STRIKE | IV BID | BID | ASK | IV ASK | DELTA |
| 72.4 | 2265 | 2346 | 76.1 | 0.64 | 22000 | 72.8 | 1156 | 1190 | 74.4 | -0.36 |
| **71.4** | **1722** | **1757** | **72.9** | **0.55** | **23000** | **71.7** | **1607** | **1641** | **73.2** | **-0.45** |
| 70.5 | 1271 | 1306 | 72 | 0.45 | 24000 | 70 | 2138 | 2184 | 72 | -0.55 |

Although there are slight rounding errors when inputting data into an options pricing calculator (24 days to expiry instead of the actual 23 days, 9 hours, along with an implied volatility entry of 71% instead of the live market IV bid of 71.4%) the options calculator was able to produce a price that is very close to the actual markets price for both the call and put (due to put-call parity) on the BTC/USD August 26th, 2022, 23000 strike options expiring in 23 days, 9hrs. As always, willing buyers and sellers set the actual market price.

| BTC/USD AUGUST 26th, 2022, 23000 CALL OPTION ANALYSIS | | | |
|---|---|---|---|
| BITCOIN PRICE | $23,093 | DIVIDEND YIELD | 0% |
| MODEL | BLACK-SCHOLES-MERTON | VOLATILITY | 71% |
| STRIKE PRICE | $23,000 | STYLE | EUROPEAN |
| TRADE DATE | Tuesday, August 2, 2022 | INTEREST RATE | 3% |
| TYPE | CALL | EXPIRATION DATE | 8/26/2022 |

| AUGUST 2nd, 2022, 23000 OPTION GREEKS | | |
|---|---|---|
| 24 DAYS TO EXPIRATION | CALL | PUT |
| PRICE | $ 1,740 | $ 1,601 |
| DELTA | 0.55 | -0.45 |
| GAMMA | 0.000094 | 0.000094 |
| THETA | -35.57 | -33.68 |
| VEGA | 23.44 | 23.44 |
| RHO | 7.19 | 7.89 |
| OMEGA | 7.29 | -6.49 |

## GREEK DYNAMICS

To finalize options pricing and the impact of the Greeks on options prices relative to changes in the reference rate, time, volatility, and interest rates, let's assume that the BTC/USD reference rate has moved higher by 1,000 points over one day from August 2nd to August 3rd, 2022, and implied volatility moved from 71% to 72%, while interest rates remained flat. With this scenario, we can calculate what the BTC/USD August 26th, 2022, 23000-strike call option market should be priced at based on Black-Scholes.

| AUGUST 2nd, 2022, 23000 OPTION GREEKS | |
|---|---|
| BTC/USD | $ 23,093 |
| 24 DAYS TO EXPIRATION | CALL |
| IMPLIED VOL | 71% |
| PRICE | $ 1,740 |
| DELTA | 0.55 |
| GAMMA | 0.000094 |
| THETA | 35.57 |
| VEGA | 23.44 |

| NEW GREEKS AFTER $1,000 MOVE OVER 1 DAY | |
|---|---|
| BTC/USD | $ 24,093 |
| 23 DAYS TO EXPIRATION | CALL |
| IMPLIED VOL | 72% |
| PRICE | $ 2,322 |
| DELTA | 0.64 |
| GAMMA | 0.000086 |
| THETA | 36.50 |
| VEGA | 22.63 |

Starting with Delta, which was 0.55 on August 2nd, 2022; as Delta represents the rate of change of an options price per $1.00 move in the BTC/USD reference rate, we can multiply the $1,000 point move by the Delta of 0.55, which is $550. Thus, the Delta impact on the options price alone is a $550 increase in value before factoring in Gamma.

Gamma is the rate of change to Delta for every $1.00 move of the BTC/USD reference rate. With a Gamma value of 0.000094 and a $1,000 point move higher to the reference rate, the impact of Gamma is calculated to be 0.094. Given this, the new Delta of the option that was at 0.55 prior to the move higher in price should be around 0.64, rounded down from 0.644. With an initial Delta of 0.55 and a final delta of 0.64, by taking the average of Delta over the one-day time series, we can calculate the net Delta impact by multiplying that average Delta of .59 by the $1,000 point move. This calculation provides an estimated net Delta impact of $595 on the price of the BTC/USD 23000-strike call option.

From our previous dataset, we see Rho is priced at $7.19, meaning for every 1% change in interest rates, either higher or lower, the option's value will increase or decrease by $7.19.

To account for the option *time decay* over one day, we take the Theta value of 35.57 and multiply it by one day, resulting in a premium decay of $35.57.

For the 1.00% move in implied volatility, from 71% to 72%, we can multiply the Vega value of 23.44 by that 1.00 point move higher in volatility, equaling an increase of $23.44.

**Initial premium + Net Delta impact - Theta + Volatility = NEW OPTION PREMIUM $1,740 + $595 - $35.57 + $23.44 = $2322.87**

| AUGUST 2nd, 2022, 23000 OPTION GREEKS | |
|---|---|
| BTC/USD | $ 23,093 |
| 24 DAYS TO EXPIRATION | CALL |
| IMPLIED VOL | 71% |
| PRICE | $ 1,740 |
| DELTA | 0.55 |
| GAMMA | 0.000094 |
| THETA | 35.57 |
| VEGA | 23.44 |

| NEW GREEKS AFTER $1,000 MOVE OVER 1 DAY | |
|---|---|
| BTC/USD | $ 24,093 |
| 23 DAYS TO EXPIRATION | CALL |
| IMPLIED VOL | 72% |
| PRICE | $ 2,322 |
| DELTA | 0.64 |
| GAMMA | 0.000086 |
| THETA | 36.50 |
| VEGA | 22.63 |

The above exercise solidifies an understanding of the impact of the major Greek variables and implied volatility on the price of an options contract when modeling for fair value. The above formula can be adjusted for different reference rates, days to expiration, implied volatility, and interest rates.

As explored above, it is best practice to enter the inputs into an options pricing calculator for exact model values rather than rely on a simplified formula with shorthand mathematics. Such simplifications for knowledge expansion cannot account for the precise changes to all the Greek values, given the dynamic relationship between each input across volatility contraction/expansion, reference price changes, and the fundamental decay from time.

CHAPTER

# PAYOFF DIAGRAMS, COLLATERAL & SETTLEMENT

## UNDERSTANDING OPTION PAYOFF FORMULAS

An options payoff formula is a function that calculates how much profit or loss an option position realizes at varying settlement prices of the underlying asset. Suppose an asset manager purchased a BTC/USD 20,000-strike call option for $3,741.00, expiring on August 26$^{th}$, 2022. What will the profit or loss be if BTC/USD settles at $25,000 at expiration?

**Inputs**:

Type: **Call**

Position: **Long**

Strike Price = **$20,000**

Premium = **$3,741.00**

BTC/USD Settlement Price Assumption = **$25,000**

**Output** = **P&L**

To calculate the profit & loss of this long call position, we first calculate the **expected intrinsic value** of the call option, which is the BTC/USD settlement price assumption minus the strike price.

$25000 - $20,000 = **$5000**

**If** the underlying asset price for BTC/USD were to settle at $25,000 on August 26, 2022, **then** the intrinsic value of the 20000-strike call option would equal $5,000 as the owner of that call option has the right to buy BTC/USD at $20,000.

Next, we must consider the premium paid to own the 20,000-strike Aug 26, 2022, BTC/USD call option to understand our *BREAKEVEN* settlement price. In this example, the premium paid for the option was $3,741. Thus, the breakeven settlement price for this long call position is **$23,741** ($20,000 + $3,741 = **$23,371).** The owner of the call option has the right to purchase BTC/USD at $20,000 but has paid $3,741 for that optionality; thus, the settlement price of BTC/USD needs to be at or above $23,741 for the position to break even or obtain a profit. **The premium paid to buy an option is the max loss for P&L calculations.** To calculate the net profit or loss for this long call position, we subtract the breakeven settlement price ($23,741) from the estimated settlement price ($25,000).

$25,000 - $23,741 = **$1,259.00**

## CRUNCHING THE NUMBERS

Given the understanding of long call option profit & loss calculations, we can calculate the P&L for the position at varying estimated settlement prices by adjusting the settlement price.

**Inputs**:

Strike Price: **20,000**

Premium: **3,741.00**

BTC/USD Settlement Price Assumption: **$25,000**

**Output** (P&L): **$1,259**

**Inputs**:

Strike Price: **$20,000**

Premium: **$3,741.00**

BTC/USD Settlement Price Assumption: **$30,000**

**Output** (P&L): **$6,259**

**Inputs**:

Strike Price: **$20,000**

Premium: **$3,741.00**

BTC/USD Settlement Price Assumption: **$20,000**

**Output** (P&L): **($3,741)**

**Inputs**:

Strike Price: **$20,000**

Premium: **$3,741.00**

BTC/USD Settlement Price Assumption: **$15,000**

**Output** (P&L): **($3,741)**

# PAYOFF DIAGRAMS

To visualize Profit & loss analysis, the same input/output function can be transported into a graph.

**Inputs**:

Strike Price: **$20,000**

Premium: **$3,741**

**Output** = **P&L**

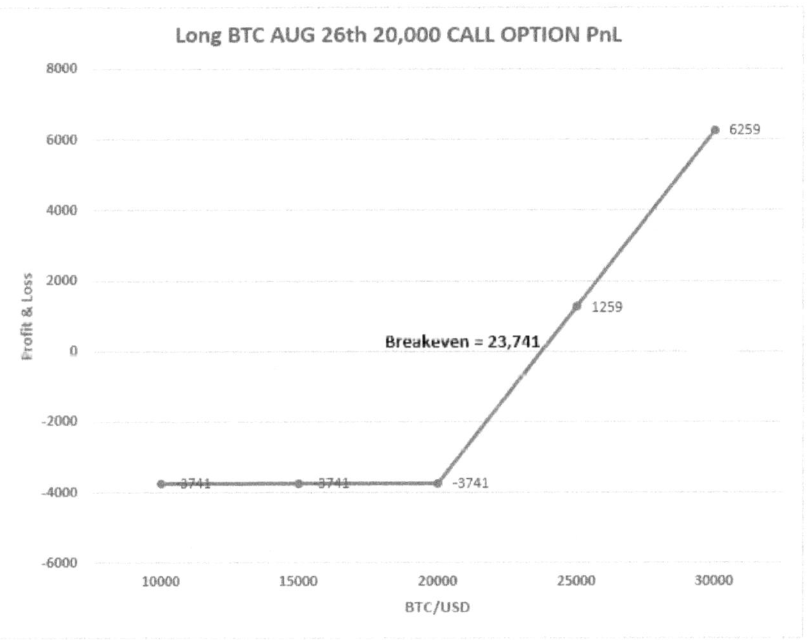


Notice that the max loss for a long call position is the premium paid ($3,741), and that loss gets smaller as BTC/USD prices move from the strike price ($20,000) to the breakeven price of 23,741, appreciating linearly into profit from the breakeven price. *When buying calls and puts, the max loss on the position is always limited to the premium paid for the option.*

## LONG PUT P&L

Since we just learned how to calculate the P&L for a long call option, we can use the same formula, except the order of operations will be adjusted to factor for the optionality to SELL the underlying asset as granted by the long-put position rather than purchase the underlying asset, as is the case for a long call option position.

Assuming a BTC/USD 20000-strike put option purchase for $589 expires on August 26th, 2022, what will the profit or loss be if BTC/USD settles at $15,000 upon expiration?

**Inputs**:

Type: **Put**

Position: **Long**

Strike Price: **$20,000**

Premium: **$589.00**

BTC/USD Settlement Price Assumption: **$15,000**

**Output** = **P&L**

To calculate the profit & loss of this long-put position, we first calculate the **expected intrinsic value** of the put option, which is the BTC/USD strike price minus the settlement price assumption.

Note: Unlike the intrinsic value calculation for a long call option, which is settlement price assumption minus strike price, the value calculation for a long-put option is reversed: strike price minus estimated settlement price.

$20,000 - $15000 = **$5,000**

**If** the price of BTC/USD were to settle at $15,000 on August 26, 2022, **then** the intrinsic value of the 20000-strike put option would equal $5,000, as the owner of that put option has the right to sell BTC/USD at $20,000.

Next, we must consider the premium paid to own the 20000 strike on Aug 26, 2022, BTC/USD put option to understand our *BREAKEVEN* settlement price. In this example, the premium paid for the option was $589.00; thus, the breakeven settlement price for this long-put position is 19,411.

$20,000 - $589.00 = **$19,411**

The owner of the put option has the right to sell BTC/USD at $20,000 and has paid $589.00 for that optionality. Thus, the settlement price of BTC/USD needs to be at or BELOW 19,411 for the position to break even or obtain a profit.

To calculate net P&L for this long-put position, we subtract the estimated settlement price ($15,000) from the breakeven price of $19,411, which is **$4,411**.

Adjusting the estimated settlement price for BTC/USD at the time of the options' expiration allows different P&L outputs to be generated and visualized on a chart.

**Inputs**:

Strike Price: **$20,000**

Premium: **$589**

BTC/USD Settlement Price Assumption: **$15,000**

**Output** (P&L): **$4,411**

**Inputs**:

Strike Price: **$20,000**

Premium: **$589**

BTC/USD Settlement Price Assumption: **$17,000**

**Output** (P&L): **$2,411**

**Inputs**:

Strike Price: **$20,000**

Premium: **$589**

BTC/USD Settlement Price Assumption: **$20,000**

**Output** (P&L): **($589)**

**Inputs**:

Strike Price: **$20,000**

Premium: **$589**

BTC/USD Settlement Price Assumption: **$25,000**

**Output** (P&L): **($589)**

# LONG PUT PAYOFF DIAGRAM

**Inputs**:

Type: **Put**

Position: **Long**

Strike Price: **$20,000**

Premium: **$589.00**

**Output**: **P&L**

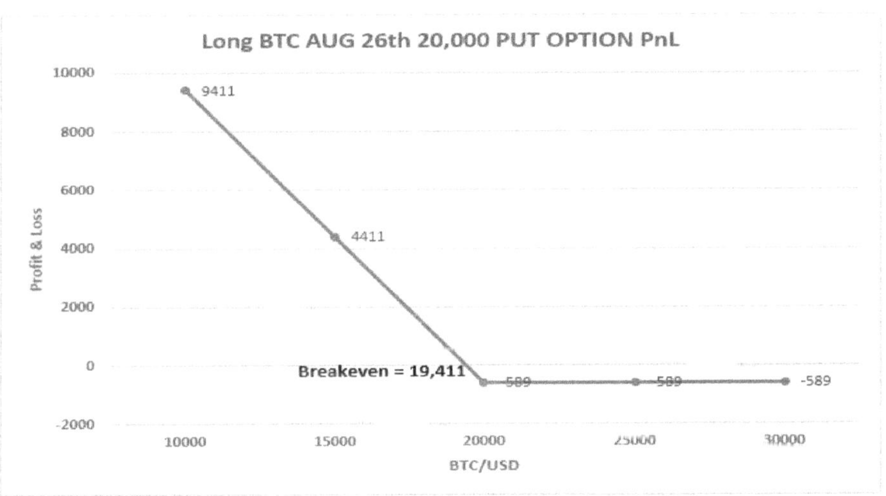


Notice the max loss of the long-put option is limited to the premium paid for the option. From the strike price of $20,000 to the breakeven price of $19,411, the option begins to accrue value, going into profit when the settlement price assumption for BTC/USD is less than the breakeven price. The profit potential is limited for put options as the underlying asset cannot trade below 0. Note: Long put option P&L calculations may differ for options on futures contracts that permit contract prices to go below zero. My Futures Markets book explains this.

## SELLING OPTIONS

Every trade in any market has a buyer and a seller. Buyers of options pay a premium to sellers for the right to exercise the option. In contrast, option sellers collect a premium from buyers and assume an obligation to fulfill the option buyer's rights. The buyer is long the option while the seller is short the option. Thus, for short P&L calculations, we take the inverse of the long P&L calculation.

For example, a trader has sold a BTC/USD 20000-strike call option for $3,741.00 that expires on August 26th, 2022. What will the profit or loss be if BTC/USD settles at $25,000 at expiration?

**Inputs**:

Type: **Call**

Position: **Short**

Strike Price: **$20,000**

Premium: **$3,741**

BTC/USD Settlement Price Assumption: **$25,000**

**Output** = **P&L**

To calculate the profit and loss of this short call position, we first calculate the *expected intrinsic value* of the call option, which is the BTC/USD settlement price assumption minus the strike price.

$25000 - $20,000 = **$5000**

**If** the underlying asset price for BTC/USD were to settle at $25,000 on August 26, 2022, **then** the intrinsic value of the 20000-strike call option would equal $5,000 as the owner of that call option has the right to buy BTC/USD at $20,000.

Next, we must consider the premium collected by the option seller, who is now obligated to sell BTC/USD at $20,000 on Aug 26th to the call buyer were the option to expire IN THE MONEY. The premium collected by the call seller is used to calculate the settlement price at which the position would *BREAKEVEN*. In this example, the premium collected for the option was $3,741. Thus, the *breakeven settlement price* for this short-call position is $23,741.

$20,000 + $3741 = **$23,741**

The call option owner has the right to purchase BTC/USD at $20,000 but has paid $3,741 for that optionality. Therefore, the settlement price of BTC/USD needs to be at or below $23,741 for the seller of the option to obtain a profit. **For option sellers, the premium collected from the sale of the option is the maximum profit**. From the strike price to the breakeven price, the profit on a short call position will diminish and turn to a loss IF the settlement price of BTC/USD is trading above the breakeven price at expiration. Given that BTC/USD can theoretically appreciate to infinity, the potential loss from selling calls is unlimited.

To calculate net profit or loss for this short call position, we subtract the estimated settlement price ($25,000) from the breakeven settlement price ($23,741):

$23,741 - $25,000 = **($1,259)**

**If** BTC/USD were to settle at $25,000 at options expiration, **then** the loss from selling the 20000-strike call option with a premium of $3,741 would be **($1,259.00)**.

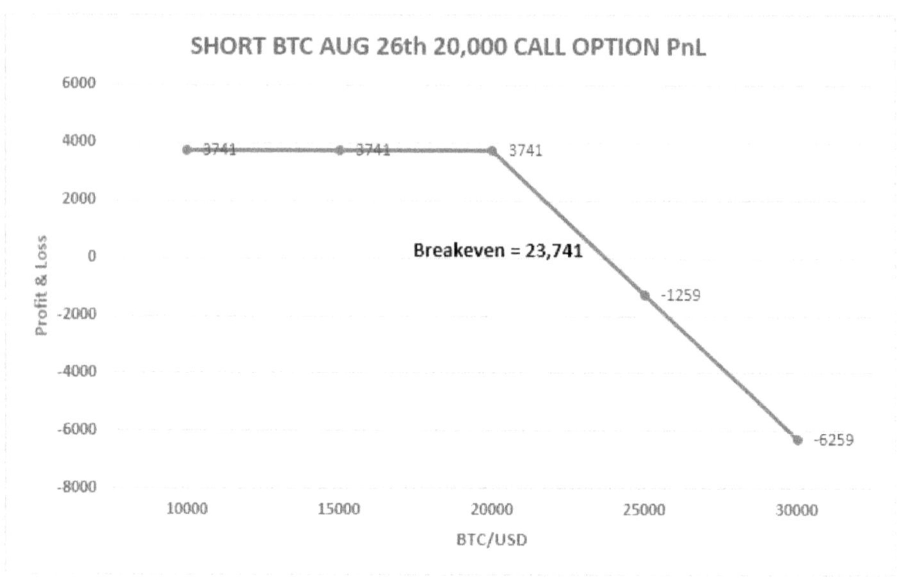

# SHORT PUT P&L

Since put option buyers have the right to sell the underlying asset at the strike price, put option sellers have an obligation to buy the underlying asset at the strike price.

For our example below, a put buyer has paid $589.00 to own the right to sell BTC/USD at $20,000 on August 26th, 2022. The seller of this option collected the $589 premium that was paid from the buyer and now must buy BTC/USD at $20,000 on August 26th, 2022, were the option to expire ITM.

The breakeven price for the put option is calculated by subtracting the premium paid for the option from the strike price:

$20,000 - $589.00 = **$19,411**

**If** BTC/USD trades above the strike price of $20,000, **then** the put seller will keep the premium paid as a maximum profit position.

**If** BTC/USD trades below the strike price, **then** the put seller's P&L will gradually erode until the breakeven price of $19,411. From there, P&L turns negative with a max loss of $19,411, were BTC/USD to settle at $0.00.

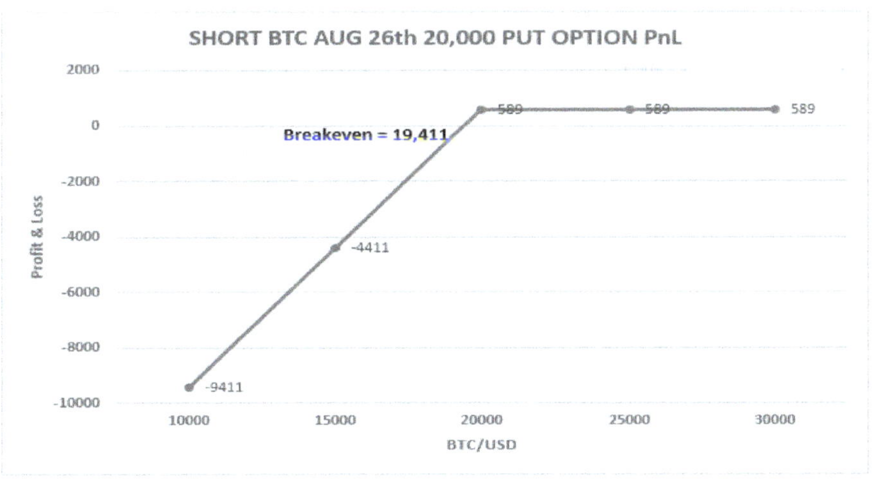

# CALL OPTIONS

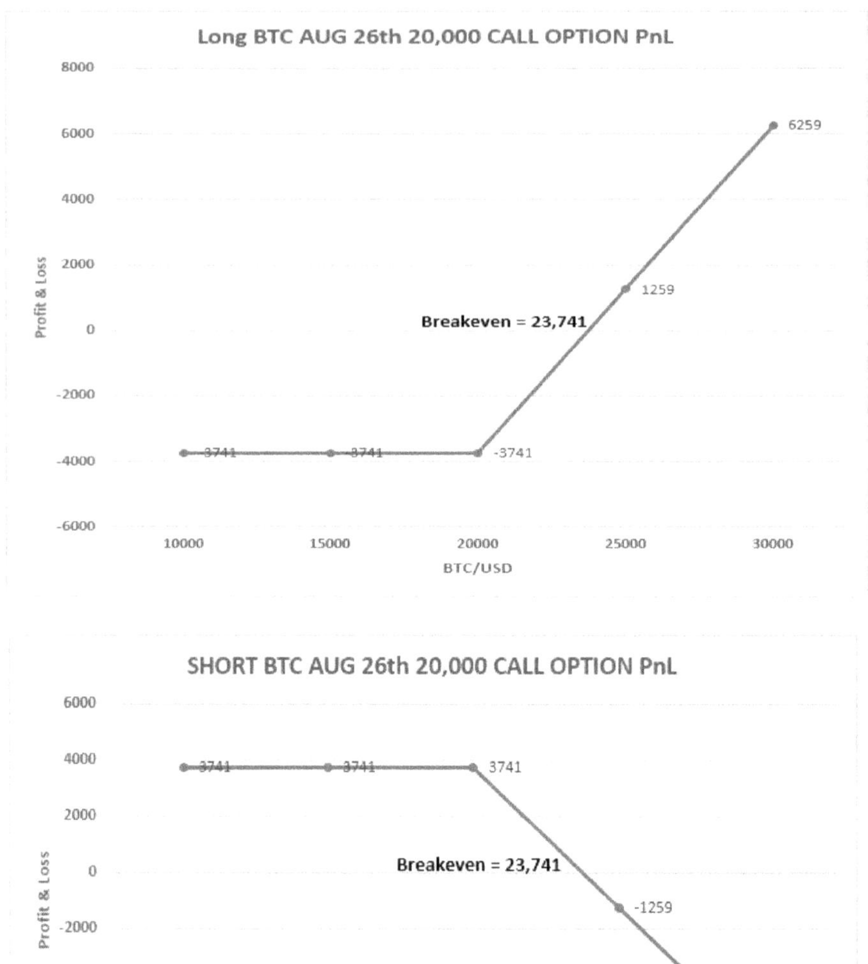

# PUT OPTIONS

Long BTC AUG 26th 20,000 PUT OPTION PnL
Profit & Loss
BTC/USD
9411
4411
Breakeven = 19,411
-589
-589
-589
10000
8000
6000
4000
2000
0
-2000
10000
15000
20000
25000
30000

SHORT BTC AUG 26th 20,000 PUT OPTION PnL
Profit & Loss
BTC/USD
Breakeven = 19,411
589
589
589
-4411
-9411
2000
0
-2000
-4000
-6000
-8000
-10000
10000
15000
20000
25000
30000

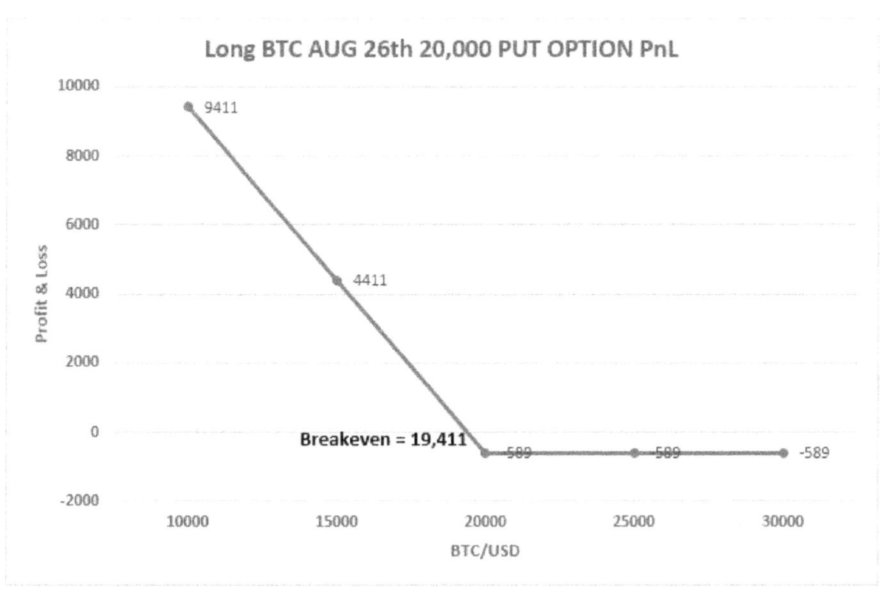

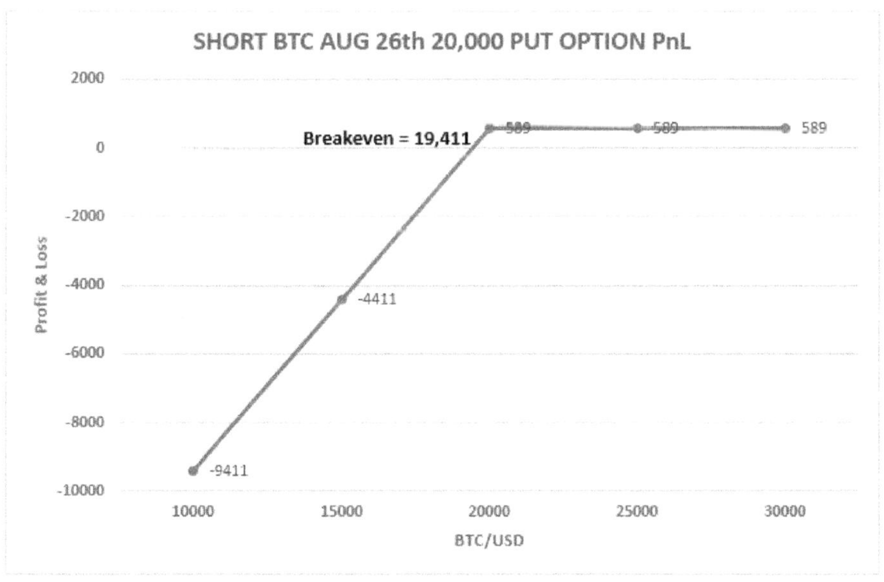

# VISUALIZING PUT-CALL PARITY & SYNTHETIC POSITIONING

In chapters one and two, we discussed the importance of *not* pricing options in a way that provides arbitrage opportunities where a trader could assemble an options portfolio that would allow for immediate profit by executing simultaneously against that portfolio in the spot or futures market. One of the examples used was the BTC/USD AUG 26, 20000 strike call option with a BTC/USD reference rate of $23,093. The call option premium in this example should not be less than $3,093; otherwise, a trader could buy the right to own BTC/USD at $20,000, pay a premium of less than $3,093, and then immediately sell BTC/USD at $23,093 and capture a spread.

PUT-CALL parity can be visualized with option payoff graphs by showing how the combination of call and put options can replicate the payoff of being long or short of the underlying asset. This ability to synthetically create a 1 Delta position further shows the importance of no arbitrage in pricing options. Below is an example of the profit & loss diagram for a long bitcoin spot market position (no options), purchased at $20,093 on Aug 2nd, 2022 ("spot" meaning actual bitcoin that exists within the Bitcoin blockchain):

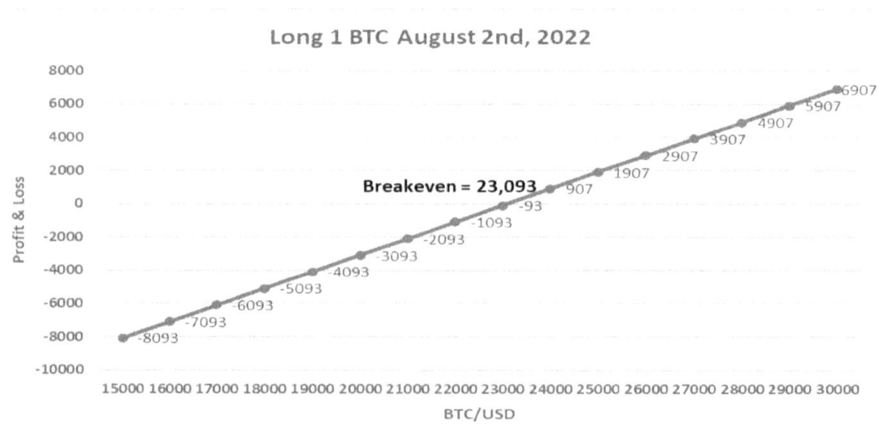

The previous chart shows the BTC/USD prices in ascending order from left to right on the x-axis and profit & loss on the y-axis. In the example, 1 bitcoin was purchased for $23,093 on August 2nd, 2022. Therefore, the position has a breakeven BTC/USD price of $23,093. At the bottom left of the chart, with a price of BTC/USD = $15,000, the position has a net loss of -$8,093. At the top right of the chart, the BTC/USD price equals $30,000; the position has a net profit of $6,907.

## SYNTHETIC LONG

Before we explore the P&L graph of a synthetic long position, assembled with AT THE MONEY options, let's refresh on the bid and offer prices for calls and puts with a 23000 strike, expiring on August 26th, 2022. Notice the asking price from a willing seller of the 23000-strike call option is offered at $1,757, and the bid from a willing buyer for the corresponding put option is $1,607.

| BTC 23,093 | | | | | AUG 26 | | | | AUG 2 | 23D 9HR |
|---|---|---|---|---|---|---|---|---|---|---|
| IV BID | BID | ASK | IV ASK | DELTA | STRIKE | IV BID | BID | ASK | IV ASK | DELTA |
| 72.4 | 2265 | 2346 | 76.1 | 0.64 | 22000 | 72.8 | 1156 | 1190 | 74.4 | -0.36 |
| 71.4 | 1722 | **1757** | 72.9 | **0.55** | **23000** | 71.7 | **1607** | 1641 | 73.2 | **-0.45** |
| 70.5 | 1271 | 1306 | 72 | 0.45 | 24000 | 70.0 | 2138 | 2184 | 72.0 | -0.55 |

Next is a P&L diagram for a LONG BTC Aug 26th, 2022, 23000-strike CALL option, and a SHORT BTC Aug 26th, 2022, 23000 PUT Option. Through the purchase of an ATM call and the simultaneous sale of an ATM put, this combination of options positions, when held simultaneously, provides a P&L profile that replicates that of a long spot bitcoin position, plus the net premium paid for the option contracts.

In this case, the CALL option that was purchased had a net debit of $1,757, and the PUT option, which was sold, provided a net credit of $1,607. Thus, this option contract portfolio of LONG 1 BTC AUG 26th 23000 CALL option and SHORT 1 BTC AUG 26th 23000 PUT option cost the position holder a net premium of **$150.**

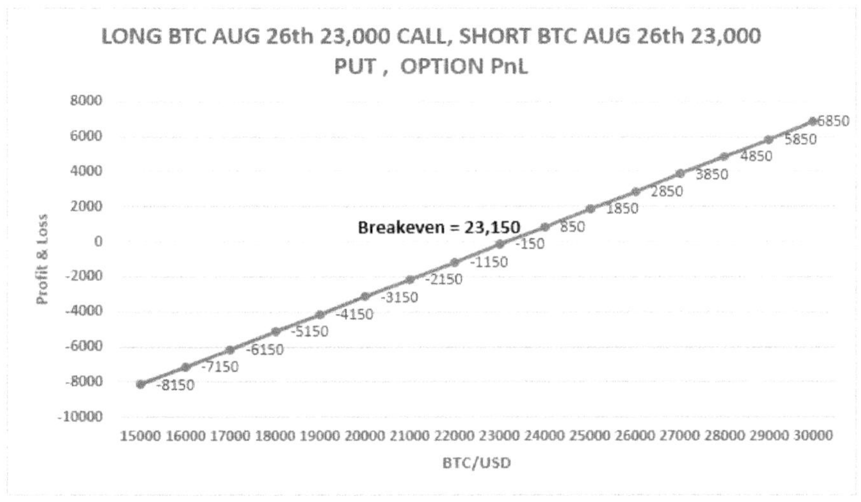


The long call option grants the right, but not the obligation, to buy bitcoin at the strike price of 23000. The sold put option obligates the seller to buy bitcoin at the strike price, replicating a long bitcoin position from $23,150 when including the net debit (premium) of the position.

Compared with the long 1 spot bitcoin payoff graph, we can visualize why there should be no arbitrage opportunities in options pricing and how options can be used to replicate a spot bitcoin position synthetically.

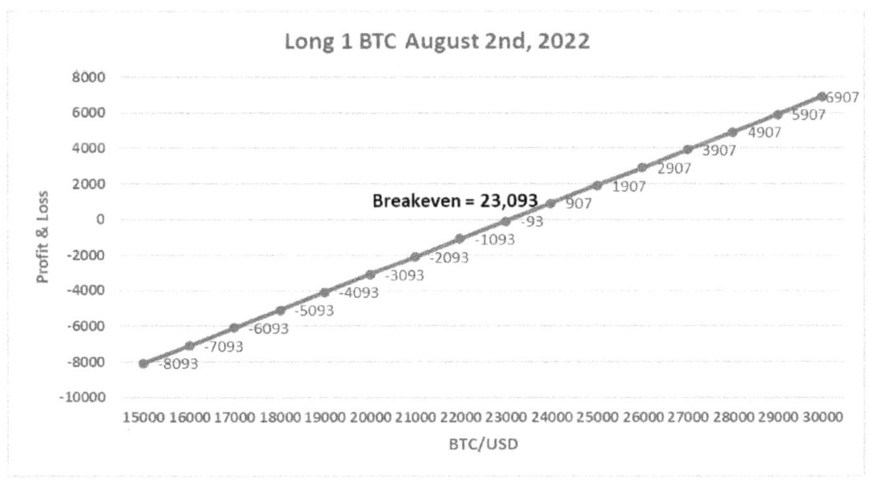

Since market participants can replicate the P&L profile of a long 1 BTC/USD spot position with options, the price paid to buy a call option and the price sold to sell a put option should never allow for arbitrage. An arbitrage scenario would allow the holder of that position to immediately borrow the underlying asset and sell it in the spot (or futures) market against the synthetic options portfolio for an immediate profit.

Think of it this way: if a trader could buy a call option and sell a put option that synthetically replicates a long spot P&L profile, the spot market price should be equal to or lower than the intrinsic value of the equivalent synthetic long option position. Otherwise, an arbitrage scenario would immediately allow the option holder to profit. It would be irrational for a market participant to knowingly give an option position away for value that provides arbitrage against spot or futures markets.

# SYNTHETIC SHORT

To hammer home the point of no arbitrage options pricing and the ability to replicate the payoff of spot or futures positions with options, below is the P&L profile of a synthetic SHORT bitcoin position:

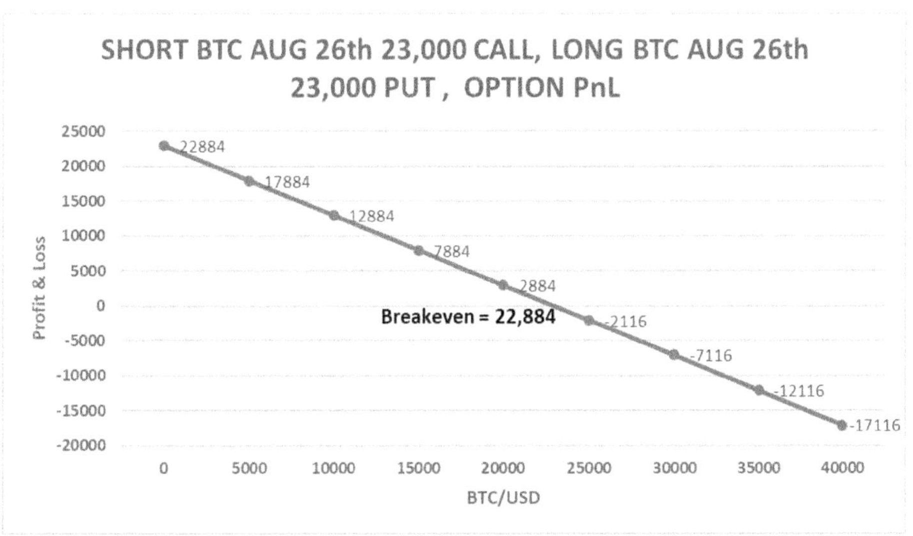


In this situation, the P&L profile of a long PUT, short CALL options position is under analysis, which provides a synthetic short position on the underlying asset. From a visual perspective, this synthetic SHORT options position provides the opposite P&L profile of a synthetic LONG position.

To construct this profile, a BTC/USD Aug 26th, 2022, 23000-strike CALL option was sold for a price of $1,641, and a BTC/USD AUG 26th, 2022, 23000-strike PUT option was purchased for a cost of $1,757, netting a debit of $116. The BTC/USD reference rate was $23,093 at trade inception.

The synthetic short option holder has bought a put option granting the right to sell bitcoin at $23,000 on Aug 26th, 2022. Simultaneously, a 23000-strike call option with the same expiration date was sold, obligating the seller to sell bitcoin at $23,000 at expiration. The reference rate at trade inception was

$23,093. Given the strike price of 23000 and the net premium paid to sell the 23000-strike call and buy the 23000-strike put of $116, the position has a breakeven price of $22,884 and begins accumulating profit IF bitcoin is trading below $22,884.

Reverting to PUT-CALL parity and the no-arbitrage approach to options pricing, it is logical to assume that a rational market would not price options to allow any participant to buy spot bitcoin and then immediately enter a synthetic short position with options that would yield a profit.

Imagine a scenario where spot bitcoin is trading for $23,093, but call and put options are priced so that one could sell a call option and buy a put option that would synthetically allow the position holder to sell bitcoin at $23,500. That scenario makes no sense and would be irrational, as any trader identifying this opportunity would buy spot bitcoin for $23,093 and sell synthetically at $23,500 until the arbitrage scenario was eliminated.

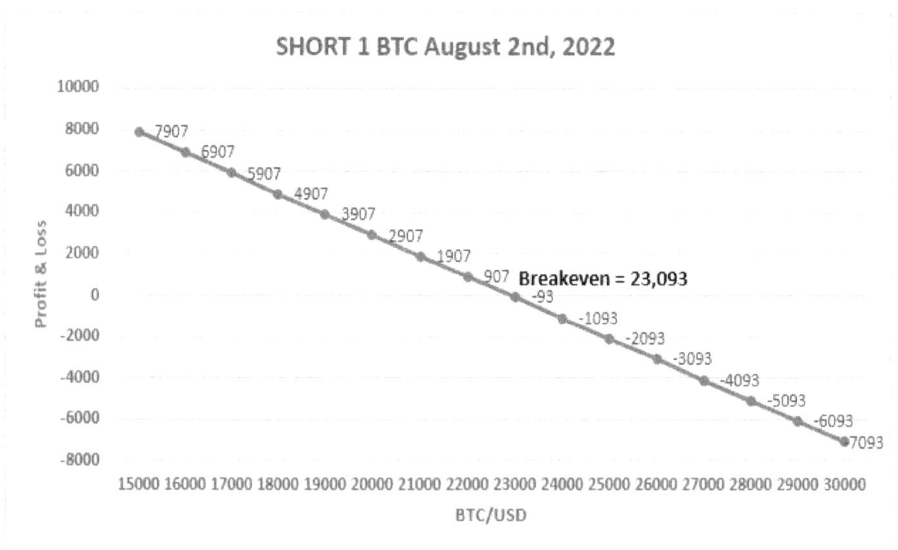

Those paying close attention will notice the synthetic short breakeven price is further from the spot breakeven than synthetic longs. This is a common occurrence in practice, as put options often trade at a premium to call options, given the psychology of seeking protection (long puts) vs. further upside exposure (long calls). This market phenomenon can be measured through a put/call relative value analysis, covered in Chapter 6, by exploring market SKEW.

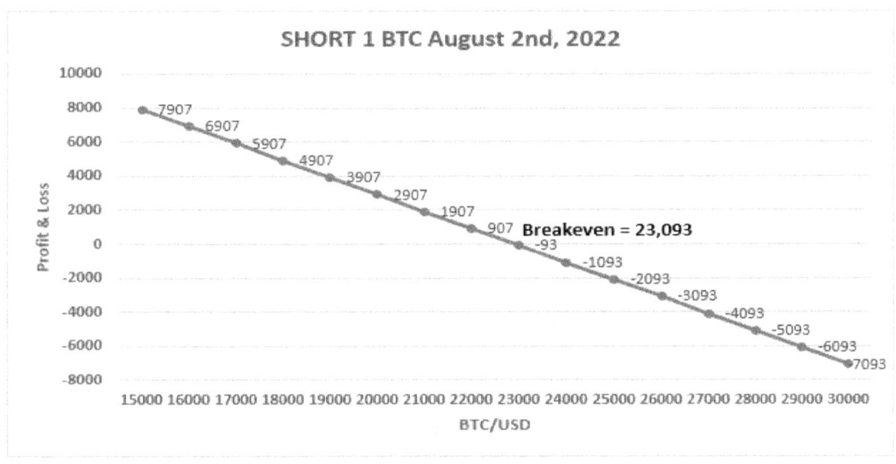


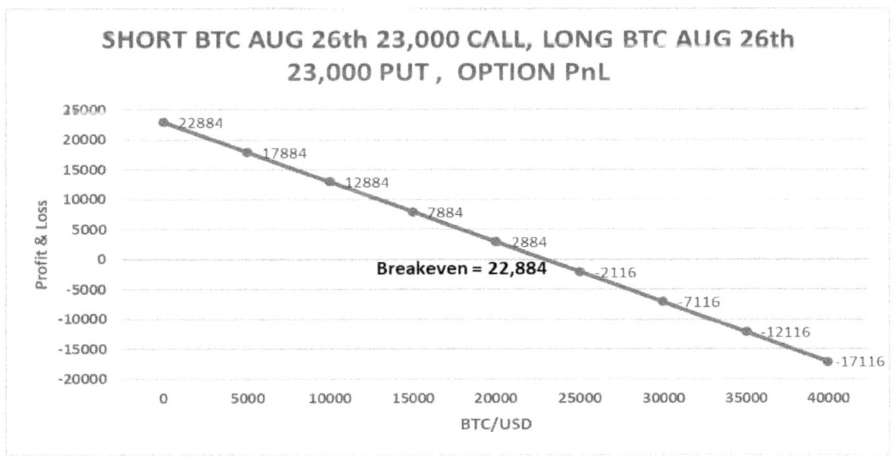

## COLLATERALIZING OPTIONS POSITIONS

The industry standard for bitcoin options requires option buyers to pay the full premium of the option upfront at the time of execution. Option sellers, on the other hand, may be permissioned to execute transactions on *margin*. Margin accounts allow for a percentage of the obligation to buy or sell bitcoin from the short option position to be held by the liquidity provider, exchange, or broker in either USD or BTC as collateral for the obligation. This collateral requirement differs from each execution venue and is a critical ruleset to define and understand when engaging in short options positions. There are two types of margin requirements: *initial margin* and *maintenance margin.*

***Initial Margin*:** The collateral must be deposited in the trading account to open a short options position at trade inception. For instance, a trading venue may require a 25% initial margin in either BTC or USD to open a short option position.

***Maintenance Margin:*** The amount of collateral that must be held in a trading account to maintain an open short option position. This requirement will differ from venue to venue.

Expanding on the initial margin, suppose a trader is long spot bitcoin and seeks to sell a bitcoin call option to generate yield against their spot bitcoin holdings. With a 25% initial margin, the trader must deposit either 0.25 bitcoin per option sold or 25% of the USD value of the BTC/USD reference rate at trade inception. When using USD to margin a short bitcoin call while the bitcoin reference rate is trading at $23,093, **then** 25% of $23,093 is $5,773.25 per option sold. This initial margin is the capital requirement needed to sell an option, subject to maintenance margin calls, were the reference price to move against the short option position. In this current example, where BTC/USD reference prices rise from $23,093 to $25,000,

25% of $25,000 equals $6,250, leading to a maintenance margin call of $476.75 for the call seller.

For maintenance margin, a trading venue will define the daily settlement time for margin calls. Bitcoin options industry standard is 8:00 UTC; thus, at 8:00 UTC, a trading venue publishes a daily settlement price for bitcoin, and this will affect the amount of collateral that must be in a trader's account to maintain any open short options positions. The trading venue's goal is to defend the integrity of the options contract, with buyers being able to exercise their rights from the obligation of options sellers. Margin requirements are a risk mitigation tool for options venues while acting as a counterparty risk mitigation tool for option market participants against the trading venue.

Suppose a trader sold short 30,000-strike call options that expire in 24 days. **If** the price of bitcoin begins to rise, let's say, to $26,500, then the likelihood of the 30000-strike call option expiring in the money increases. Correspondingly, so does the obligation of the short call seller to sell bitcoin at the strike price or cash settle the difference between strike and settlement price, were the option to expire ITM.

Therefore, the trading venue may define the maintenance margin for short call options as 25% of the daily bitcoin settlement price if margining in USD rather than margining in the obligatory asset for a short bitcoin call seller – bitcoin, where the margin requirement would remain fixed at 0.25 bitcoin until the option became ITM, triggering a margin call.

For short call and put options, trading venues may also view the maintenance margin as dynamic, meaning that from the initial margin requirement, the maintenance margin percentage requirement may increase with the probability that the option sold short will expire ITM. Continuing with our short call example, where a trader decided to sell 30000-strike bitcoin call options that expire in 24 days with a reference rate of $23,093; were the market to rise to $26,500, the maintenance margin may increase from 25% to

33% or 50%, depending on the rules of the trading venue. Since the short call position must sell bitcoin at $30,000, the trading venue manages the risk of the option seller's ability to meet the obligation or offset the position by purchasing the option to close the short position. Conversely, suppose the bitcoin price falls from $23,093 to $20,000. In that case, the trading venue may return a percentage of collateral to the call option seller, given the decline in the probability that the 30000-strike option will expire ITM.

The same methodology can be used to understand initial margin and maintenance margin requirements for put options. Put option sellers are obligated to buy bitcoin at the strike price, so collateralizing the position with USD makes sense, although venues may still allow collateralization with bitcoin. If the reference price for BTC/USD is $23,093 and a trader seeks to sell a 20000-strike put option that expires in 24 days, with an initial margin requirement of 25%, the put option seller must deposit either 0.25 bitcoin or $5,773.25.

If bitcoin prices move lower, the maintenance margin percentage requirement for a short put option may increase as the reference rate of bitcoin decreases since the probability of put options expiring ITM increases as the reference rate of the underlying asset moves lower. Given this, a move from $23,093 to $22,000, or 4.7% lower, could increase the maintenance margin from the initial 25% (0.25 BTC or $5,773.25) to 29.7% of the reference price, or $6,534, forcing the short put holder to deposit $760.75 or an additional 0.047 bitcoin.

# SETTLEMENT

Bitcoin is a global market, accessible 24/7 each day of the year. Daily settlement prices are disseminated on UTC time for options, with the industry standard being 8:00 UTC. Every day at 8:00 UTC, the settlement price for bitcoin is published and options positions will be "marked to market." Net profit and loss will be calculated on all positions to determine if a margin call needs to be initiated from options sellers or returned to option sellers in rebalancing with the maintenance margin requirements.

When an option meets its expiration date, the reference rate posted on that date will be the settlement reference rate in determining whether an option has expired in or out of the money. Based on that calculation, the net profit or loss on the position will be realized. Examine the BTC/USD August 26th, 2022, options below at the expiration date with a settlement reference price of **$23,100**.

- ❖ ***Intrinsic value***: Options with intrinsic value at expiration are *IN THE MONEY.*
- ❖ ***Extrinsic value***: Extrinsic value equals zero at expiration, and those options are *OUT OF THE MONEY.*

| BTC/USD 23,100 | | | | | 26 Aug, 2022 | | | | 26-Aug | 0D 0HR |
|---|---|---|---|---|---|---|---|---|---|---|
| | CALL | | | | | | | PUT | | |
| IV BID | BID | ASK | IV ASK | DELTA | STRIKE | IV BID | BID | ASK | IV ASK | DELTA |
| | 5100 | | | 1 | 18000 | | | | | |
| | 4100 | | | 1 | 19000 | | | | | |
| | 3100 | | | 1 | 20000 | | | | | |
| | 2100 | | | 1 | 21000 | | | | | |
| | 1100 | | | 1 | 22000 | | | | | |
| | **100** | | | **1** | **23000** | | | | | |
| | | | | | 24000 | | 900 | | | 1 |
| | | | | | 25000 | | 1900 | | | 1 |
| | | | | | 26000 | | 2900 | | | 1 |
| | | | | | 27000 | | 3900 | | | 1 |
| | | | | | 28000 | | 4900 | | | 1 |
| | | | | | 30000 | | 6900 | | | 1 |

## PHYSICAL VS. CASH SETTLEMENT

Depending on trading venue rules, at options expiration, a trader may be able to choose between *cash settlement* and *physical settlement*, as defined below:

***Physical Settlement*:** Call & put options that expire ITM are physically exercised, obligating call sellers to deliver bitcoin to the call buyer and put sellers to buy bitcoin from the put buyer at the respective strike price.

***Cash Settlement***: Call & put options that expire ITM with the seller's obligation fulfilled through a USD (or bitcoin equivalent) payment that covers the contract's notional value difference between the strike price and the settlement price.

Suppose a market participant sold short 10 BTC/USD on August 26th, 2022, 20000 call options. At the expiration date, the settlement price for BTC/USD came in at $23,100. To physically settle the short call obligation, this market participant can sell 10 bitcoin at an effective price of $20,000 to physically settle the contract, receiving $200,000. To cash settle the position, the trader can subtract the strike price from the settlement price ($23,100 - $20,000), which is $3,100. With ten contracts sold short, the obligation can be cash-settled by depositing $31,000 or 1.34 bitcoin. Cash settlement avoids the trader having to sell all ten bitcoin at $20,000.

In both the physical and cash settlement scenarios, the net notional value of the call seller's obligation of $31,000 or 1.34 bitcoin has been covered, and the call option buyer's optionality has been fulfilled. For physical settlement, the call option seller deposited 10 bitcoin to the trading venue and received $200,000 in return, which is a $31,000 difference from the return she would have received by selling 10 bitcoin at the settlement reference price of $23,100 were she not to have been obligated to sell at $20,000 from the short call position. In the cash settlement scenario, the trader opted to deposit the net notional value of the obligation, which was either $31,000 or 1.34 bitcoin (spot reference $23,100).

CHAPTER

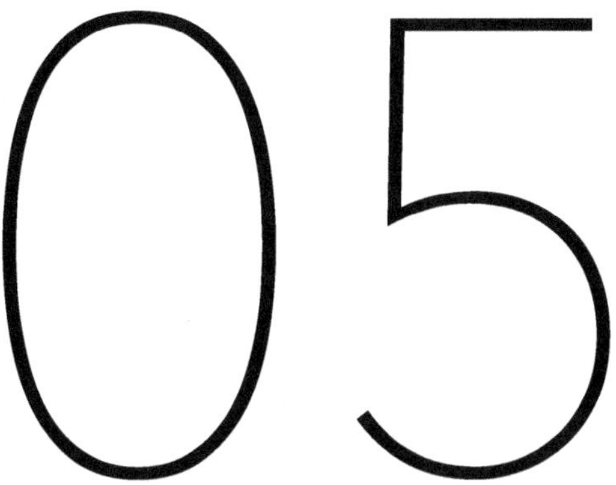

# STRATEGY & EXECUTION

## HOLD ON FOR DEAR LIFE

Those who have been around bitcoin long enough will undoubtedly have heard or seen the acronym "HODL." The meaning behind this phrase about bitcoin centers around the asset's price volatility. Elevated price volatility is expected during these early stages of information distribution, adoption, and real-world use for bitcoin as electronic cash and a store of value asset. The term stems from the conviction of bitcoin HODL'rs to hold the asset through these wide price swings.

HODL'rs take a long view on the ability of bitcoin to fundamentally change how people view energy transfer and the compensation mechanisms for work, along with the hope for increased general awareness of inflation-resistant personal energy storage. Bitcoin HODL'rs: individuals, hedge funds, family offices, bitcoin miners, private cities, or nation states may seek to *optimize* their bitcoin holdings. Optimization can be accomplished by deploying a variable percentage of bitcoin holdings into risk management and fundamental yield strategies with options markets, just as precious metals HODL'rs may do.

# HODL YIELD

The first strategy to explore is the HODL YIELD, more commonly known as a **COVERED CALL OPTION.** This simple strategy generates yield from the basics of call optionality. Bitcoin owners can sell OTM call options, which obligate them to sell their bitcoin at higher prices than the current market. In return for this sale obligation, a call seller receives a premium payment from the call buyer.

Call options are the right to own an underlying asset at a negotiated price (strike price) during a predefined time (expiration date). Sellers of call options are obligated to sell the underlying asset at the strike price if the underlying asset's settlement price were to be at or above the strike price at expiry. The **HODL YIELD** strategy seeks to do just that – sell call options that are fully collateralized or *covered* by bitcoin holdings.

Sticking with the August 26th, 2022, options market, suppose a HODL'r is happy to hold bitcoin long-term, with a time frame analysis in years or decades rather than days, weeks, or months. Suppose a HODL'r has read this book_and has decided to apply some of that knowledge of options in practice by selling OTM call options against 10% of their bitcoin holdings. After analyzing the market and being armed with Delta knowledge, the HODL'r has decided to define a strategy where, every month, they will sell 20 Delta call options, fully collateralized with bitcoin.

With fully collateralized short call options, the risk of the position is that bitcoin trades above the strike price at expiration, obliging the call seller to sell bitcoin at a lower price than the price at the time of options expiration. This HODL'r is willing to accept that opportunity cost against 10% of their bitcoin holdings in exchange for the opportunity to generate passive income while *holding on for dear life* through volatile price discovery during the adoption phase of bitcoin globally.

| BTC 23,093 | | | | | AUG 26 | | | | AUG 2 | 23D 9HR |
|---|---|---|---|---|---|---|---|---|---|---|
| **IV BID** | **BID** | **ASK** | **IV ASK** | **DELTA** | **STRIKE** | **IV BID** | **BID** | **ASK** | **IV ASK** | **DELTA** |
| 74.1 | 3630 | 3741 | 81.1 | 0.80 | 20000 | 77.1 | 554 | 589 | 79.2 | -0.20 |
| 72.5 | 2890 | 3006 | 78.4 | 0.73 | 21000 | 74.6 | 809 | 843 | 76.4 | -0.28 |
| 72.4 | 2265 | 2346 | 76.1 | 0.64 | 22000 | 72.8 | 1156 | 1190 | 74.4 | -0.36 |
| **71.4** | **1722** | **1757** | **72.9** | **0.55** | **23000** | **71.7** | **1607** | **1641** | **73.2** | **-0.45** |
| 70.5 | 1271 | 1306 | 72 | 0.45 | 24000 | 70.0 | 2138 | 2184 | 72.0 | -0.55 |
| 69.1 | 901 | 947 | 71.3 | 0.36 | 25000 | - | - | 4647 | 150 | -0.64 |
| 68.8 | 635 | 670 | 70.5 | 0.28 | 26000 | - | - | 5318 | 150 | -0.72 |
| 68.5 | 439 | 462 | 69.9 | 0.21 | 27000 | - | - | 6024 | 150 | -0.79 |

Analyzing the market, the HODL'r can identify the strike price closest to 20 Delta for call options, which in this scenario is the 27000-strike call option that expires in 23 days, 9 hours. Refreshing on Delta, which is the rate of change for the options price relative to a $1.00 move in bitcoin reference prices, either up or down, the HODL'r understands Delta can also be used for shorthand math in calculating the probability that the market is pricing for the option to expire in the money. With a Delta of 0.21 on August 2$^{nd}$, 2022, the options market is pricing in a 21% chance that bitcoin will be trading at or above $27,000 in 23 days, 9 hours at expiry on August 26$^{th}$, 2022.

Given the HODL'r holds 10 bitcoin and has decided to deploy 10% of their holdings into the HODL YIELD strategy, they will seek to sell 1 BTC/USD August 26, 2022, 27000-strike CALL option to the buyer with the best bid. The best bid is currently $439 or 0.019 bitcoin (0.019 x $23,093 = $439). The premium earned can be collected in either USD or in bitcoin. Given the HODL'rs view on bitcoin relative to USD, they may choose to receive the premium in bitcoin.

# CALCULATING ANNUALIZED YIELD

There are two ways to calculate annualized yield when selling BTC/USD covered calls depending on how the premium is collected – either in bitcoin or U.S.A. dollars.

When the yield generated from the premium received by selling a bitcoin-covered call option is collected in U.S.A. dollars, the formula is as follows:

**USD Option Premium / Bitcoin Reference Rate * (365 / Days to Expiration)**

$439.00 / $23,093 * (365/24) =

0.019 x 15.20 = **29 % annualized rate of return**

If the yield generated from the premium received by selling a bitcoin-covered call option is collected in bitcoin, the formula is:

**Bitcoin Option Premium * (365 / Days to Expiration)**

0.019 bitcoin * 15.20 = **29% annualized rate of return.**

Working with the HODL'rs venue of choice that requires a 25% initial margin deposited as collateral for the short options position, the HODL'r will deposit 0.25 bitcoin into their trading account and then sell 1 BTC/USD August 26$^{th}$, 2022, 27000-strike CALL option, whereafter the 0.019 bitcoin will be credited to the HODL'rs trading account.

At expiration, **if** the bitcoin settlement price is below $27,000, **then** the 27000-strike call option that was sold will expire worthless, and the HODL'r will keep the premium credited to the HODL'rs account. **If** the bitcoin settlement price exceeds $27,000, **then** the HODL'r has two options: sell 1 bitcoin for $27,000 OR settle the net notional value difference between the strike price and the settlement price.

Avenue #1 is straightforward; sellers of call options are obligated to sell the asset at the strike price if the option expires in the money. However, if the call option seller does not want to sell the asset, they can opt for avenue #2 and ***cash settle*** the account relative to the difference between strike price and settlement price.

With bitcoin trading at $30,000, the option seller would settle 0.1 bitcoin or $3,000 with the trading venue to cover the obligation from the short call option. This approach is known as ***cash settlement***, the notional value equivalent of physical settlement. Trading venues will have different settlement procedures for physical and cash-settled options. It is always best practice to be in tune with the settlement methodology for any trading venue.

To further grasp *cash settlement* vs. physical settlement, think of it this way – was the option to be physically settled, meaning the call option seller chose to sell one bitcoin at $27,000 per the contractual obligation, the call option seller would receive $27,000 from selling one bitcoin to the call option buyer. With that cash, the call option seller could immediately repurchase bitcoin at the settlement price of $30,000, a net difference of $3,000. The HODL'r would have sold one bitcoin for $27,000 and immediately bought $27,000 worth of bitcoin at the settlement price of $30,000, which would be 0.9 bitcoin.

By choosing cash settlement, the call option seller does not sell one bitcoin at $27,000. Instead, the position is settled by sending 0.1 bitcoin to the HODL'rs trading venue to cover the net obligation of the short call position that settled in the money at $30,000. In both scenarios, the net loss from the position is the same: 0.1 bitcoin or $3,000 before accounting for the credit received from selling the option.

The HODL'r received 0.019 bitcoin from selling the August 26th, 2022, 27000-strike call option from the call buyer (worth $438.76 when BTC/USD was at $23,093 during trade execution). With a net loss or opportunity cost from selling the option equaling $3,000 or 0.10 bitcoin, by adding the premium collected from the option sale of $438.76 or 0.019 bitcoin, **the *net loss* of the position equals $2,561.24 or 0.081 bitcoin.**

The net loss from the position can be viewed as an opportunity cost since the price of bitcoin moved significantly higher within the life of the option (23d, 9hr) from a reference rate of $23,093 on August 2nd to a theoretical settlement price on August 26th of $30,000. Given that the HODL'r is hedging 10% of their net bitcoin position, which is one bitcoin, the net opportunity cost of $2,561.24 or 0.081 bitcoin can be considered covered by the total portfolio gains.

At the option sale trade date, the HODL'r held ten bitcoin and sold 1 BTC/USD call option expiring in 23d, 9hr, collecting $438.76 or 0.019 bitcoin with the BTC/USD reference rate at $23,093. The 10-bitcoin held had a net USD notional value of $230,930. During options expiration on August 26th, bitcoin's theoretical settlement reference rate was $30,000, causing the 27000-strike call seller to lose 0.081 bitcoin or $2,561.24. However, the HODL'r now owns 9.919 bitcoin with USD conversion rates at $30,000 for a net USD notional value of $297,570. If the HODL'r did not hedge with options, the portfolio would be worth $300,000 with BTC/USD conversion rates at $30,000. However, bitcoin prices easily could have moved lower or settled below the option's strike price at expiration.

***Physical Settlement***: You can fulfill the obligation of the short option by selling the asset at the strike price (short call option) or buying the asset at the strike price (short put option).

***Cash Settlement***: This method fulfills the short option's obligation by covering the net difference between the strike price and the settlement price.

# HODL YIELD PAYOFF DIAGRAM

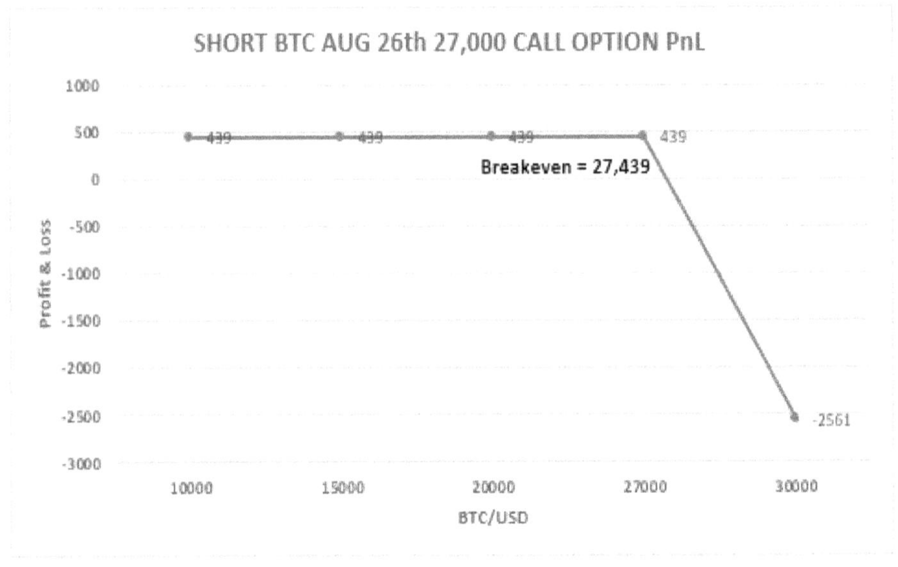

## NO FREE LUNCH

Although the best things in life may be free, when discussing risk management and real yield strategies, there is always an opportunity cost in exchange for generating passive income or purchasing insurance as risk management for the underlying asset. That opportunity cost must be weighed against the alternative of doing nothing and letting your portfolio free float with the market through time.

In the example scenario above, the bitcoin settlement price was $30,000. The HODL'r sold the 27000-strike call option, collecting 0.019 bitcoin as a premium for passive income. The ***net opportunity cost*** in bitcoin equaled 0.081 bitcoin, as the cash settlement requirement was a $3,000 difference between the strike and settlement price at expiration (or 0.1 bitcoin.) 0.1 bitcoin minus 0.019 bitcoin received as the premium yield from the option sale equals **0.081 bitcoin**. In this case, the call seller opted to receive their premium in bitcoin instead of USD (0.019 bitcoin instead of $439.00), saving $131.00 at expiration. These savings are from the 0.019 bitcoin received as a premium payment, increasing in value from $439.00 to $570.00 with a bitcoin settlement price of $30,000. This difference in premium value can be identified on the payoff graph from the prior page, where the payoff graph assumes the premium received in USD.

The HODL YIELD short call example may have faced an opportunity cost, but the HODL'r is still likely very happy with the outcome. Bitcoin's price has increased from $23,093 to $30,000 in 23 days and 9 hours, resulting in a gain of $6,907 per bitcoin, or a 30% increase. The call seller had allocated 10% of their net holdings to the strategy, which left them with 9.919 bitcoin from their original 10 bitcoin with a net gain of $66,640.

The most common question about the HODL YIELD strategy, otherwise known as a covered call, is: What should a market participant do if the settlement price was greater than the strike price of the call option sold at the time of expiration (as in the scenario above)? This question advances strategy and execution tactics for our next trade setup, *the FOMO YIELD.*

## FOMO YIELD

Another acronym bitcoin observers and market participants are likely to have heard is **FOMO – FEAR OF MISSING OUT**. The term derives from the feeling of watching bitcoin prices move higher without proper exposure to the asset from spot holdings or derivatives positions. For many, FOMO comes from the sense of missing an opportunity. These misses are either due to a miscalculation in expected price movements or sudden, unexpected market events that drive price or volatility in the anticipated direction without owning the proper position to benefit.

To solve FOMO, participants can utilize the options market to define a strategy that suits these criteria. In the HODL YIELD strategy, a covered call option position was explored, where a market participant is long-spot bitcoin and seeks to generate passive income from the sale of call options. For the **FOMO YIELD** strategy, we will mimic the approach of the HODL YIELD, but instead of selling covered call options, the FOMO YIELD strategy sells *cash-secured put* options. When selling a put option, the obligation is to buy the asset at the strike price IF the settlement price of bitcoin at option expiration were trading below the strike price. As with a covered call, where the obligation is collateralized with bitcoin (or USD equivalent), put options can be fully collateralized or covered by holding USD (or the bitcoin equivalent) through expiry.

Using a similar rule set to the HODL YIELD strategy above, suppose a market participant looking to purchase bitcoin is faced with the question of market timing. Rather than buying spot bitcoin today, a HODL'r can sell 20 Delta put options monthly until they are obligated to buy the asset at the strike price. After some analysis, the HODL'r has identified the BTC/USD August 26th, 2022, 20000-strike put option currently pricing with a 20 Delta. The best bid for the option is $554.00 or 0.024 bitcoin. Applying the yield formula, selling the 20,000-strike put option would generate a 36.5% yield, and the strike price is 13.3% below the current market price of $23,093.

Since the obligation is to buy bitcoin, the HOLD'r has decided to post USD collateral into their trading account as a margin. With a 25% initial margin requirement to sell the 20000-strike put option, they deposit 25% of the current BTC/USD reference rate, fulfilling the requirement to sell the 20000-strike put option. ($23,093 x 0.25 = **$5,773.25**). With the initial margin deposited and one option contract sold, this market participant has opted to receive their premium in USD, providing their trading account with a net credit of $554.00. If the settlement price of bitcoin at expiration were above the strike price of $20,000, they would keep the entire premium as passive income. If bitcoin settlement prices at expiration came in below the strike price, the market participant would have to either take physical delivery and buy one bitcoin at $20,000 or cash settle the difference between the strike price and the settlement price at expiration.

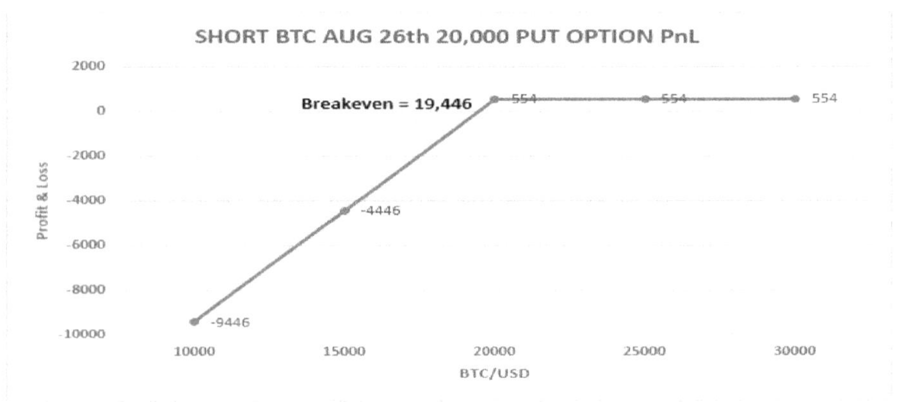


Let's assume that the settlement price for bitcoin at expiration came in at $18,000, a decrease of 22% from $23,093, where bitcoin was priced when the put option was sold. To physically settle the position, the market participant must buy one bitcoin at $20,000, even though the settlement price of bitcoin is $18,000. Given their obligation to the 20,000-strike put buyer who has the right to sell bitcoin at $20,000, the physical settlement assigns one bitcoin to the put seller's trading account and debits $20,000. With bitcoin trading at $18,000, the position had an opportunity cost of $2,000.

To cash settle the position, the put seller can subtract the settlement price ($18,000) from the strike price ($20,000), which is -$2,000. Physical settlement and cash settlement are equivalent net notional settlement mechanisms. However, physical settlement results in actual bitcoin ownership for the put seller, whereas cash settlement leaves the put seller with a net debit against their trading account. With cash settlement, the opportunity cost would be realized by simply debiting $2,000 dollars from the put seller's account, leaving the account with a USD balance of $18,000 instead of a long 1 bitcoin balance as with physical settlement.

The opportunity cost of $2,000 for the short 20000-strike put options with a bitcoin settlement price at expiration of $18,000 had a net opportunity cost of **-$1,446** when factoring in the premium yield of $554.00 from selling the 20000-strike put option. Had the market participant opted to buy the spot market at $23,093 instead of selling the 20000-strike put option, that cost would have been **$5,093** ($23,093 - $18,000).

## HODL YIELD MEET FOMO YIELD

Revisiting the final question from the **HODL YIELD** strategy section: what can a market participant do if the settlement price of bitcoin at expiration is higher than the strike price of the call option sold? This scenario obligates physical or cash settlement of the ITM option. When a market participant sells a covered call, ITM short-call obligations can present an opportunity to deploy the **FOMO YIELD** strategy into the following month's options market.

From the HODL YIELD strategy example, a HODL'r sold the BTC/USD August 26th, 2022, 27000-strike call options, and the bitcoin settlement price at expiration came in at $30,000. This caused an opportunity cost for passive income to be $3,000 gross. **If** the market participant opts for physical settlement, their account will now have a $27,000 balance from selling one bitcoin at $27,000. Therefore, they can utilize the **FOMO YIELD** strategy by selling a *cash-secured put* option at the same strike they sold the *covered call* option in the prior month.

Forecasting options prices, we can take the *known knowns* for the next month's 27000-strike put option expiring on September 30$^{th}$, 2022, by plugging in variables into a Black-Scholes-Merton options pricing calculator:

Strike Price: **27000**

Reference Rate: **$30,000**

Type: **Put**

Style: **European**

Position: **Short**

Trade Execution Date: **August 26$^{th}$, 2022**

Settlement Date: **September 30th, 2022**

Interest Rate: **4.00%**

Implied Volatility Assumption: **70%**

With these inputs, the options pricing calculator returns a value for the 27000-strike put option, expiring on September 30th, 2022, of **$1,237 or 0.041 bitcoin**.

Suppose a market participant who sells covered calls and gets called away at expiration seeks to buy back bitcoin at the strike price they were called away from. In that case, they can sell a put option at that same strike price, generating passive income in exchange for the obligation to buy back bitcoin at that strike price.

Please note that markets are random, and bitcoin may not trade lower into the short put strike during the following month(s). Suppose bitcoin were to continue trading above the short put strike. In that case, the put seller will not be obligated to buy bitcoin and will only collect a premium as passive income from the put option sale, as the option will expire worthless if bitcoin is trading above the put strike at expiry.

This scenario presents an opportunity to sell the 27000-strike put options until the market trades back below $27,000 during options expiration. Then, the market participant buys back their bitcoin, initially sold by deploying the **HODL YIELD** strategy that sold the 27000-strike call option. Once bitcoin is repurchased from a short put obligation, the HODL YIELD strategy could again be deployed by selling covered calls in the following month's options market.

In the following graph, the **FOMO YIELD** payoff for selling the BTC/USD September 30th, 2022, 27000-strike put option is shown:

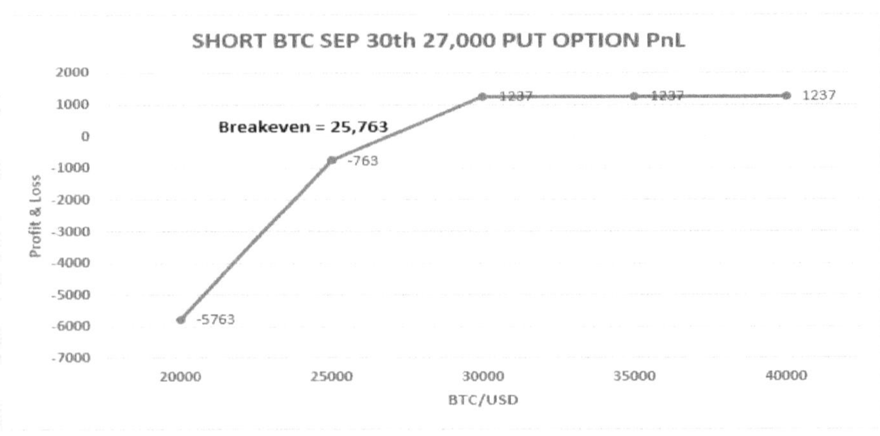


**If** the short put seller decides to receive the premium yield in USD, then that annualized rate of return could be calculated as $1,237/$30,000 x (365/36) = 0.041 x 10.14 = **42% annualized yield**.

If the put seller opts to receive the premium yield in bitcoin, they will receive **0.041 bitcoin** per option sold.

## OPTIMIZED YIELD

The next strategy to discuss, dubbed the **OPTIMIZED YIELD**, combines the HODL YIELD & FOMO YIELD strategies and is designed for market participants with bitcoin and USD in their portfolio. The strategy goes one step further and caps losses from the short option positions by purchasing further OTM call and put options, limiting downside exposure to the purchased option strikes.

Selling call options generates passive income, as does selling put options, with risk minimized to opportunity cost, when the short positions are fully collateralized and capped at the strike prices of the purchased OTM options. For example, when selling call & put options, a fully collateralized position means the market participant has one bitcoin (or the USD equivalent) per option sold, deposited at their trading venue or accessible in their bitcoin wallet.

On August 2nd, 2022, the bitcoin reference rate was trading at $23,093. An August 26th, 2022, 20000-strike put option was sold for $554.00, while the 27000-strike call option was sold for $439.00. The 33000-strike call option was also purchased for $49.00, and the 14000-strike put was purchased for $2.79.

In legacy financial markets, this strategy is known as a short-covered strangle. A *strangle* is an options position that executes the same position (long or short) in both calls and puts with the same expiration date at different strike prices. To calculate the annualized yield of this position, a trader can sum the net premium from the call & put option sales ($439 + $554 = **$993**), subtract the premium paid for purchased options ($51.79), then divide by the bitcoin reference rate at time of execution, multiplied by (365/DTE): $941.21/$23,093 x (365/24) = 0.041 x 15.20 = **62% annualized yield.**

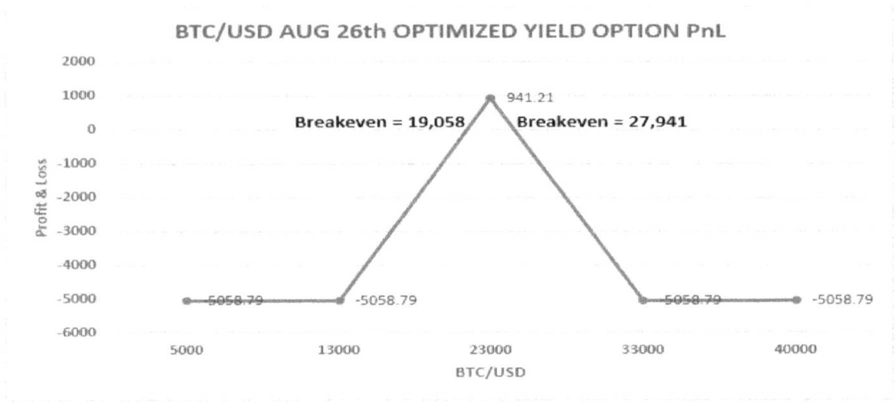


This position obligates the sale of bitcoin to be at $27,000 and the purchase of bitcoin to be at $20,000. It also grants the right to purchase bitcoin at $33,000 and sell bitcoin at $14,000, which caps the potential opportunity cost of the short options. In exchange for the obligation, this position generates 0.041 bitcoin or $941.21, providing a 62% annualized rate of return.

# OPTIMIZED HODL

Those that are long bitcoin can build upon the **HODL YIELD** strategy, where a bitcoin-covered call is sold to generate passive income through yield generation from premium earned by selling call options. For the **OPTIMIZED HODL** strategy, we will forgo yield and instead utilize that premium collected from selling call options to purchase put options for ***ZERO COST***.

Long PUT options are insurance on where a long bitcoin position can be sold since put option owners have the right to sell bitcoin at the strike price. To pay for such insurance, bitcoin holders can sell upside call options, with the premium collected to fund the insurance cost fully. Any residual premium can be delegated as passive income after the insurance purchase.

Sticking with the August 26th, 2022, options market, the **OPTIMIZED HODL** strategy will be constructed by selling the same call option as in the HODL YIELD example; the BTC/USD August 26th, 2022, 27000-strike call option, which is paying $439.00 or 0.019 bitcoin. For the **OPTIMIZED HODL** strategy, instead of collecting the premium earned from selling the call option as yield, we will analyze the put option market and identify which put option we can buy at zero cost with the premium earned from the call option sale.

| BTC 23,093 | | | | | AUG 26 | | | | AUG 2 | 23D 9HR |
|---|---|---|---|---|---|---|---|---|---|---|
| **IV BID** | **BID** | **ASK** | **IV ASK** | **DELTA** | **STRIKE** | **IV BID** | **BID** | **ASK** | **IV ASK** | **DELTA** |
| | | | | | 18000 | 84.1 | 254 | 277 | 86.3 | -0.10 |
| - | 3759 | 5147 | 121 | 0.86 | **19000** | **79.7** | **369** | **393** | **81.5** | **-0.15** |
| 74.1 | 3630 | 3741 | 81.1 | 0.80 | 20000 | 77.1 | 554 | 589 | 79.2 | -0.20 |
| 72.5 | 2890 | 3006 | 78.4 | 0.73 | 21000 | 74.6 | 809 | 843 | 76.4 | -0.28 |
| 72.4 | 2265 | 2346 | 76.1 | 0.64 | 22000 | 72.8 | 1156 | 1190 | 74.4 | -0.36 |
| **71.4** | **1722** | **1757** | **72.9** | **0.55** | **23000** | **71.7** | **1607** | **1641** | **73.2** | **-0.45** |
| 70.5 | 1271 | 1306 | 72 | 0.45 | 24000 | 70.0 | 2138 | 2184 | 72.0 | -0.55 |
| 69.1 | 901 | 947 | 71.3 | 0.36 | 25000 | - | - | 4647 | 150 | -0.64 |
| 68.8 | 635 | 670 | 70.5 | 0.28 | 26000 | - | - | 5318 | 150 | -0.72 |
| **68.5** | **439** | **462** | **69.9** | **0.21** | **27000** | - | - | 6024 | 150 | -0.79 |
| 68.6 | 300 | 323 | 70.2 | 0.16 | 28000 | 65.1 | 5132 | 5294 | 76.3 | -0.84 |

Highlighted in bold are the at-the-money options for the 23000-strike calls and puts, the 27000-strike call option market, and the subsequent put option market that can be bought for zero cost from selling the 27000-strike call options. In simple terms, this position obligates a market participant to sell bitcoin at $27,000 while providing the right to sell bitcoin at $19,000.

The net premium breakdown is as follows:

SHORT 27000-strike CALL: +$439.00 (or 0.019 bitcoin)

LONG 19000-strike PUT: -$393.00 (or 0.017 bitcoin)

**Residual Premium: $46.00 (or 0.002 bitcoin), 3% annualized yield**

The OPTIMIZED HODL strategy is like a synthetic short options position, except it is not executing against ATM options but OTM options, which provide upside exposure to the short call strike in return for the insurance to sell bitcoin at the put strike.

In other words, the **OPTIMIZED HODL** is a defensive strategy for risk management that allows for participation in upside price appreciation, capped at the call strike + premium, which in this example is **17% higher** than the current reference price of bitcoin at $23,093.

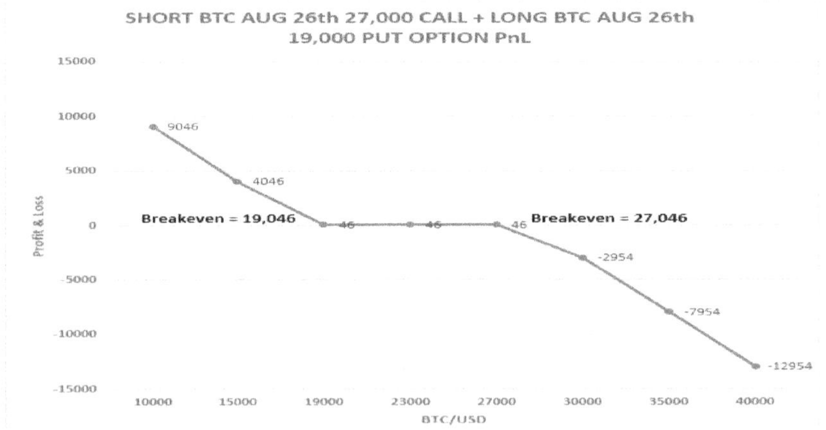

## OPTIMIZED FOMO

Just as the OPTIMIZED HODL strategy builds on the HODL YIELD, so will the **OPTIMIZED FOMO** strategy build on the FOMO YIELD. The OPTIMIZED HODL & FOMO strategies encompass purchasing and selling call-and-pull options with different strike prices and the same expiration date. In legacy finance, such an assembly of options are known as *collars*.

With the **OPTIMIZED FOMO,** market participants can *use the premium earned from selling put options to buy call options* at **ZERO COST**. The purchase of call options grants the right to buy bitcoin at the call strike, with the premium earned from the short put obligation to buy bitcoin at the put strike.

From the above example, the BTC/USD August 26th, 2022, 20000-strike put option was paying $554.00 or 0.024 bitcoin, annualizing a 36.5% rate of return. Analyzing the BTC/USD August 26th, 2022, call option market, participants can identify which call option could be purchased at zero cost with the premium earned from the put sale. The August 2nd, 2022, market snapshot shows that the BTC/USD August 26th, 2022, 27000-strike call option is offered at $462.00 or 0.020 bitcoin.

The net premium breakdown is as follows:

SHORT 20000-strike PUT: +$554.00 (or 0.024 bitcoin)

LONG 27000-strike CALL: -$462 (or 0.020 bitcoin)

**Residual Premium: $92.00 (or 0.004 bitcoin), 6% annualized yield.**

The FOMO YIELD is an excellent strategy for those seeking exposure to bitcoin yet want to avoid buying bitcoin at current market prices. The short put position obligates the put seller to buy bitcoin at a lower price than the current market price and generates a yield that can be received as bitcoin or USD. Those with major FOMO but not enough to buy spot bitcoin today may be willing to forgo yield generation from FOMO YIELD and instead deploy the **OPTIMIZED FOMO** strategy. The latter approach provides the potential to profit from higher bitcoin prices with the long call position financed with the premium earned from the short put sale.

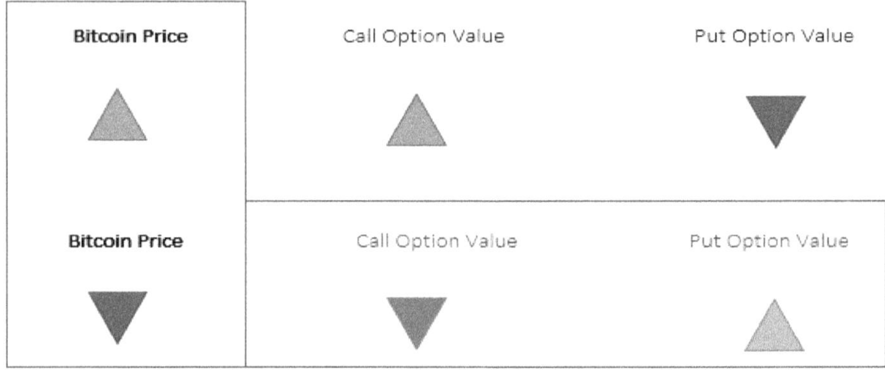


The **OPTIMIZED FOMO** strategy is like a synthetic long options position, except it is not executing against ATM options but OTM options. The **OPTIMIZED FOMO** provides upside exposure at the call strike in return for the obligation to buy bitcoin at the put strike.

In other words, the **OPTIMIZED FOMO** is a defensive strategy for managing the risk of a portfolio that does not have the desired allocation to bitcoin that the market participant views as ideal. This approach allows for participation in upside price appreciation above the call strike and the obligation to buy bitcoin at a **13.3% discount** to the current market price of $23,093.

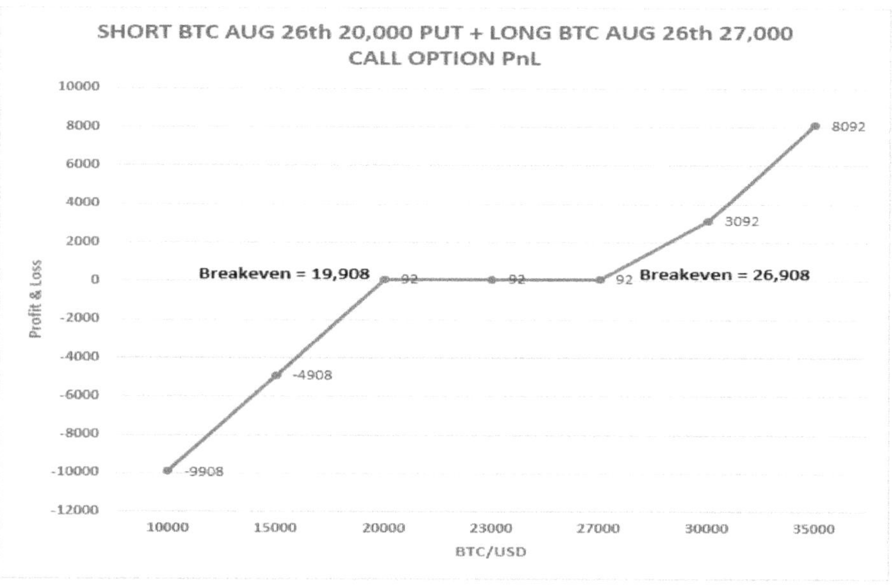

## VOLATILITY STRATEGIES

Bitcoin options are multivariable, deriving value not just from the price of bitcoin but also from implied volatility, time to expiration, and interest rates. So far, we have learned option strategies focused on the bitcoin reference price variable, but we can also derive strategies for the following major variable in options pricing - *VOLATILITY.*

Although technically tradable, interest rates do not have nearly as much effect on options prices compared to volatility. When selling options for passive income, market participants are already trading time, as the closer to expiration an option gets, the faster an option loses its time value or extrinsic value.

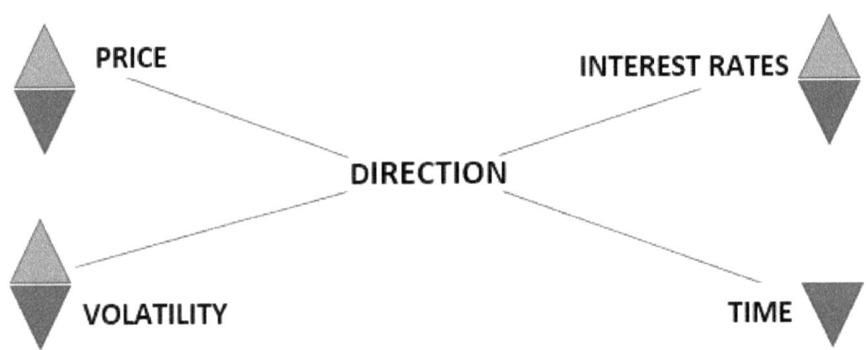


There are two main types of volatility: historical (realized) and implied:

***Historical (realized) Volatility*** is the annualized standard deviation of lognormal distribution from past daily asset price movements.

***Implied Volatility*** **(IV)** is a forward-looking and dynamic variable that estimates the anticipated future standard deviation of lognormal price distributions. A bid and offer price determines IV from willing buyers and sellers. *Implied volatility is tradable in real-time.*

To review, for future bitcoin price range estimates, standard deviation states that:

One standard deviation represents the price range that bitcoin should trade in roughly 68% of the time.

Two standard deviations represent the price range bitcoin should trade in roughly 95% of the time.

Three standard deviations represent the price range in which bitcoin should trade in roughly 99% of the time.

Bitcoin is an example of a highly volatile asset, given that BTC/USD encompasses wide price swings and ranges across multiple time horizons. In other words, bitcoin has a high standard deviation – the price range that bitcoin will trade in 68%, 95%, and 99% of the time is broader than that of a low standard deviation asset.

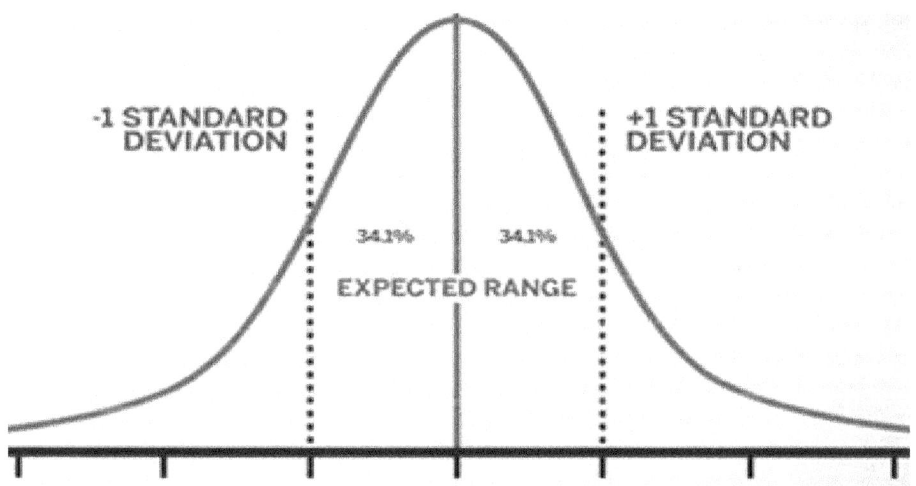


*The price range for a one standard deviation move expands for assets with high implied volatility levels. In comparison,* a lower implied volatility level *tightens* the range of prices for a one-standard-deviation move.

When IV changes over time, the price of options and the expected range of the underlying asset from trade inception until options expiration dramatically differ from an implied volatility of 10% to 100%.

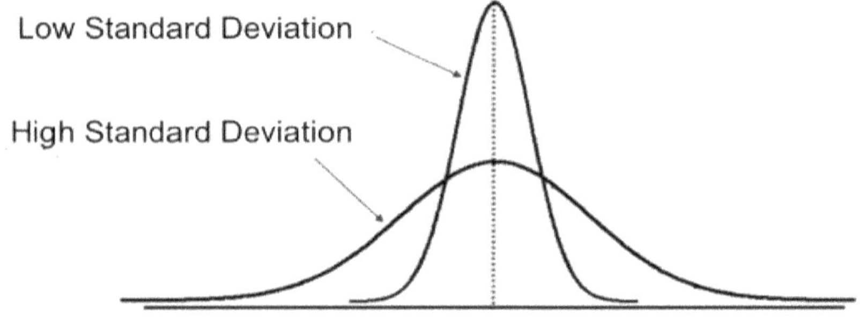


[1]

High Volatility: Wide price swings and ranges.

Low Volatility: Narrow price swings and ranges.

Referencing the BTC/USD AUG 26th, 2022, 23000 -strike call option market in the table below, the current market price is a buyer willing to bid $1,722 for the call option with an implied volatility of 71.4%. This means the expected *annualized* price range of the underlying asset is 71.4% of the bitcoin reference rate, $23,093 ($6,605 - $39,581), with a one standard deviation price range for the bitcoin reference rate within 24 days until option expiration on Aug 26th, 2022, being +/- $4,228 ($18,865 - $27,321) from the current price of $23,093.

| BTC/USD | BTC AUG 26TH, 2022 23000 CALL | | |
|---|---|---|---|
| 23,093 | IV | PRICE | EXPECTED RANGE |
| | 100 | $2,411 | $17,172 - $29,014 |
| | **71.4** | **$1,722** | **$18,865 - $27,321** |
| | 50 | $1,239 | $20,133 - $26,053 |
| | 30 | $769 | $21,317 - $24,869 |
| | 10 | $302 | $22,501 - $23,685 |

As demonstrated in the tables above and on the next page, all else equal, if the IMPLIED VOLATILITY of an underlying asset falls, call and put options decrease in value, and the expected price range for BTC/USD narrows. All else being equal, if the VOLATILITY of an underlying asset rises, call and put options increase in value, and the expected price range for BTC/USD widens. Given this, we can construct options strategies that gain or lose value depending on the direction of *volatility*.

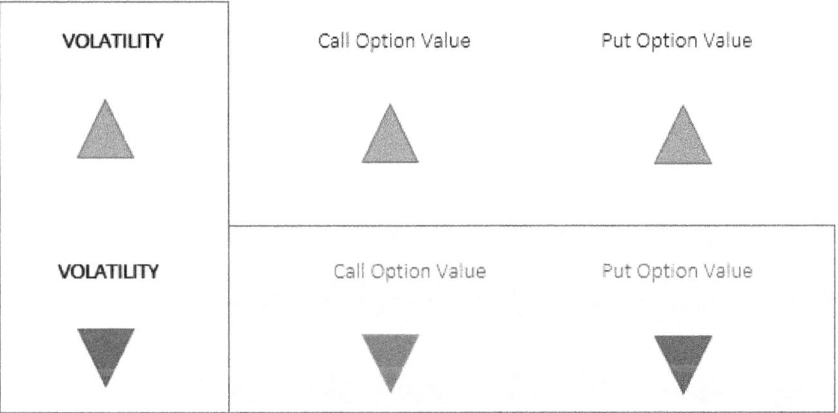


When trading volatility strategies, the Delta and Gamma of an option must be considered. Delta is the rate of change to the value of an option for every $1.00 move in the underlying asset. Gamma is the rate of change for Delta, which increases as the reference price of an underlying asset approaches the strike price.

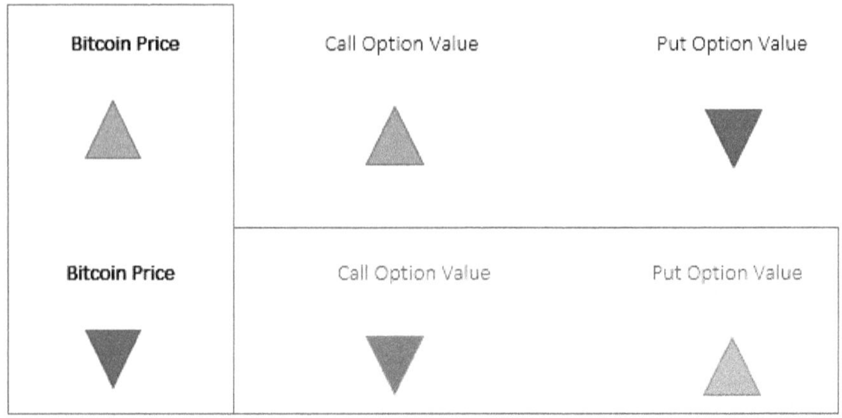

Like a standard bitcoin price chart that shows historical price data points on a line graph over a selected time frame, volatility data can also be aggregated and displayed on a line graph. Below is a historical volatility chart for bitcoin from mid-2017 to August 14th, 2023.

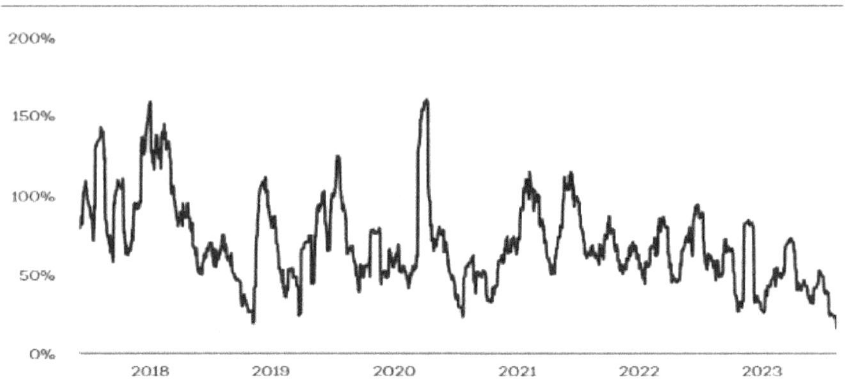

## FOMO VOL (LONG STRADDLE)

Suppose a market participant wanted portfolio exposure to increasing bitcoin volatility; how can options be assembled where the payoff graph of the position creates profit scenarios where volatility rises? Introducing the **FOMO VOL** strategy, which by legacy definition is a *long straddle* position. The **FOMO VOL** strategy is designed around ATM options, such as buying both a call and put with the same strike price closest to the current BTC/USD reference rate.

Since the **FOMO VOL** strategy buys options, the max loss of the position is the premium paid to purchase ATM call & put options. Looking at the August 26th, 2022, market with a BTC/USD reference rate of $23,093, the 23000-strike call and put options are at the money. The contracts are offered at $1,757.00 and $1,641.00, respectively. Buying both these options requires a market participant to pay a $3,398.00 net premium.

| BTC 23,093 | | | | | AUG 26 | | | | AUG 2 | 23D 9HR |
|---|---|---|---|---|---|---|---|---|---|---|
| IV BID | BID | ASK | IV ASK | DELTA | STRIKE | IV BID | BID | ASK | IV ASK | DELTA |
| 72.4 | 2265 | 2346 | 76.1 | 0.64 | 22000 | 72.8 | 1156 | 1190 | 74.4 | -0.36 |
| **71.4** | **1722** | **1757** | **72.9** | **0.55** | **23000** | **71.7** | **1607** | **1641** | **73.2** | **-0.45** |
| 70.5 | 1271 | 1306 | 72 | 0.45 | 24000 | 70.0 | 2138 | 2184 | 72.0 | -0.55 |

This strategy would be deployed by a market participant anticipating volatility to increase, thus causing an increase in options prices given the expanded expected range of a one standard deviation move for the BTC/USD reference rate. Anticipated or unanticipated volatility moves are event-driven by some force that motivates the purchase or sale of assets, weighted toward the relative unexpectedness. The more unexpected the event, the wider the expected ranges associated with explosive moves and the higher or lower the reference rate can be. Understanding implied volatility's impact on options prices could present an opportunity to explore strategies that define profit & loss scenarios from the expansion and contraction of the expected range.

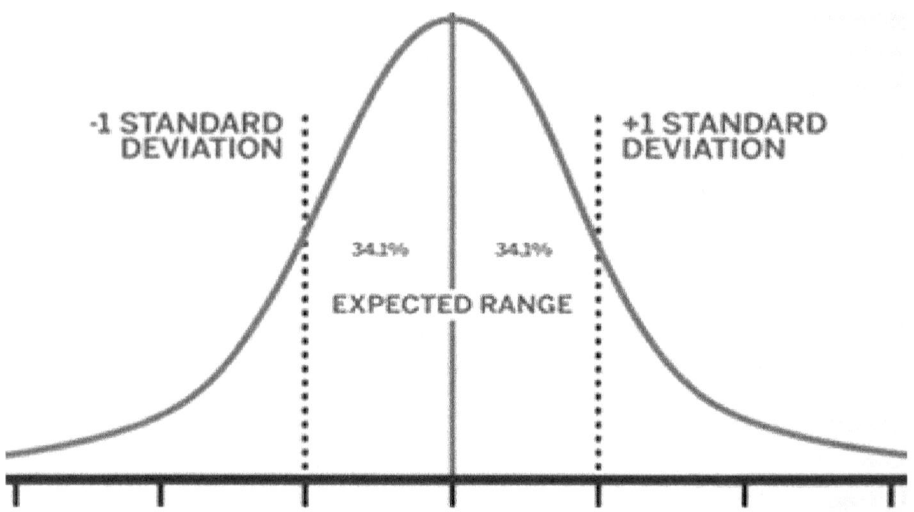


As the strategy requires a net premium outlay to be paid up-front, we can add the net premium of $3,398 to the 23000-strike call option. This aids in understanding the breakeven price and the upside for the strategy. Subtract the net premium from the 23000-strike put option to determine the downside breakeven price. This can be visualized below with the **FOMO VOL** strategy payoff graph for the August 26th, 2022, options market.

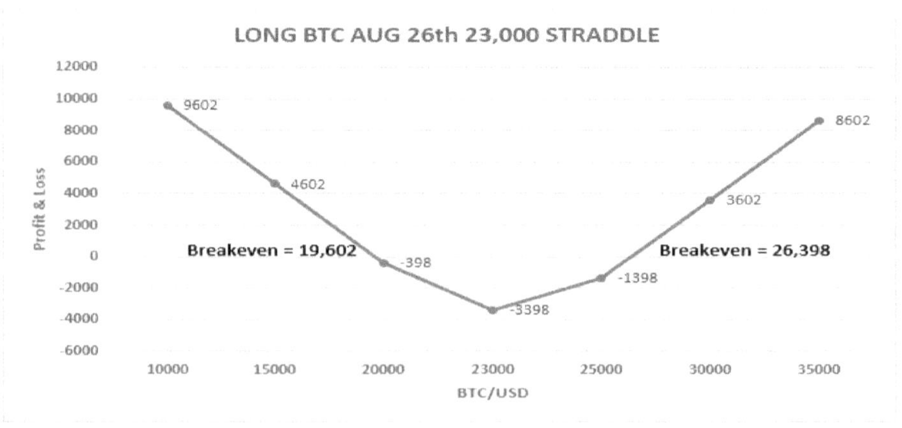

Visualizing how long volatility strategies can profit from the expansion in the expected range helps assess the cost to buy or sell that optionality.

The straddle range can be measured against the expected implied volatility range when determining relative value. With BTC/USD reference prices at $23,093, the ATM straddle, or FOMO VOL strategy example, has a downside breakeven price of $19,602 and an upside breakeven price of $26,398 compared to the expected ranges for the option given different levels of implied volatility.

| BTC/USD | BTC AUG 26TH, 2022 23000 CALL | | |
|---|---|---|---|
| 23,093 | IV | PRICE | EXPECTED RANGE |
| | 100 | $2,411 | $17,172 - $29,014 |
| | **71.4** | **$1,722** | **$18,865 - $27,321** |
| | 50 | $1,239 | $20,133 - $26,053 |
| | 30 | $769 | $21,317 - $24,869 |
| | 10 | $302 | $22,501 - $23,685 |

# OPTIMIZED VOL (COVERED SHORT STRADDLE + LONG WINGS)

The **OPTIMIZED VOL** strategy is very similar to the OPTIMIZED YIELD strategy, except the **OPTIMIZED VOL** strategy has a tighter range in obligatory strike prices, albeit a larger yield generated from the option sales.

The strategy is constructed by selling at-the-money call and put options and collecting the net premium as passive income. The OPTIMIZED VOL is designed for market participants who hold bitcoin and USD to cover the obligation from the call and put option sales.

| BTC 23,093 | | | | | AUG 26 | | | | AUG 2 | 23D 9HR |
|---|---|---|---|---|---|---|---|---|---|---|
| **IV BID** | **BID** | **ASK** | **IV ASK** | **DELTA** | **STRIKE** | **IV BID** | **BID** | **ASK** | **IV ASK** | **DELTA** |
| 72.4 | 2265 | 2346 | 76.1 | 0.64 | 22000 | 72.8 | 1156 | 1190 | 74.4 | -0.36 |
| **71.4** | **1722** | **1757** | **72.9** | **0.55** | **23000** | **71.7** | **1607** | **1641** | **73.2** | **-0.45** |
| 70.5 | 1271 | 1306 | 72 | 0.45 | 24000 | 70.0 | 2138 | 2184 | 72.0 | -0.55 |

The at-the-money call and put options are bid for $1,722 and $1,607, respectively. Selling these options aggregates a net premium of $3,329. The max loss of the short options positions is capped by including the purchase of OTM call and put options, limiting the upside and downside losses to the strike price of the purchased options. The 33000-strike call option can be purchased for $49.00, and the 13000-strike put can be purchased for $2.00. When we include the total cost of $51.00 to purchase these options, the net premium earned from this structure is $3,278.

To annualize the yield, take the net premium earned, divide it by the current bitcoin reference price, and multiply by (365/DTE): $3,278 / 23,093 x 15.20 = **215% annualized yield.**

Although the annualized yield may look attractive, this strategy pays well because it has a high opportunity cost. The OPTIMIZED VOL is designed to profit from range-bound bitcoin markets experiencing diminishing levels of volatility. Historically, bitcoin has been a highly volatile asset.

The payoff graph for this example of the OPTIMIZED VOL is shown below. Note that the strategy is profitable while the BTC/USD reference rate remains between the lower and upper bounds of the breakeven prices. Losses build quickly as volatility expands and reference prices move outside the breakeven range. The losses are capped from the long OTM put & call options purchased at the 13000 strike and 33000 strike.

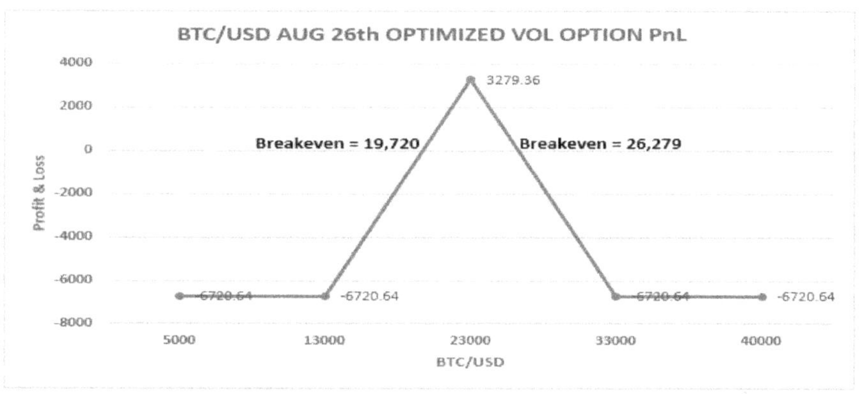


Compare the **OPTIMIZED VOL** strategy payoff graph with the OPTIMIZED YIELD strategy payoff graph:

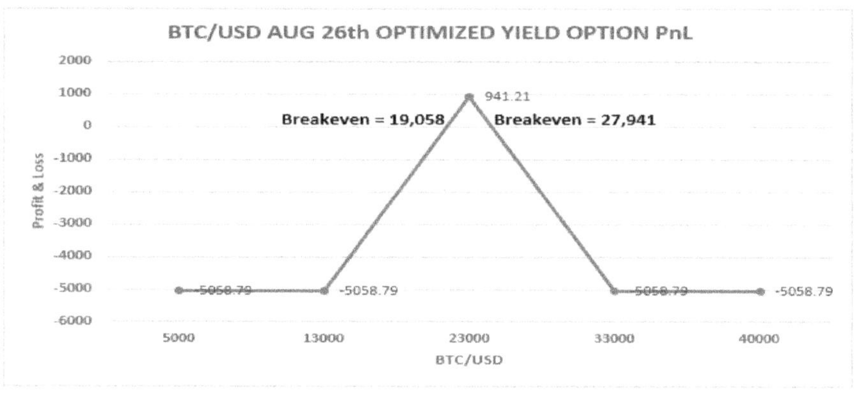

## VOLSHOT (LONG STRANGLE)

Playing off the word longshot, the VOLSHOT strategy is a net debit to a trader's account, meaning they need to pay for the options up front to put on the position. The maximum loss of the position is the net premium paid for the options. The VOLSHOT strategy is designed to profit from expanded volatility and is constructed by purchasing out-of-the-money calls and put options.

The structure shown below will consist of a long BTC/USD August 26th, 2022, 27000-strike call option and a long August 26th, 2022, 19000-strike put option, with an aggregate premium paid of $855.

| BTC 23,093 | | | | | AUG 26 | | | | AUG 2 | 23D 9HR |
|---|---|---|---|---|---|---|---|---|---|---|
| **IV BID** | **BID** | **ASK** | **IV ASK** | **DELTA** | **STRIKE** | **IV BID** | **BID** | **ASK** | **IV ASK** | **DELTA** |
| - | 3759 | 5147 | 121 | 0.86 | 19000 | 79.7 | 369 | 393 | 81.5 | -0.15 |
| 74.1 | 3630 | 3741 | 81.1 | 0.80 | 20000 | 77.1 | 554 | 589 | 79.2 | -0.20 |
| 72.5 | 2890 | 3006 | 78.4 | 0.73 | 21000 | 74.6 | 809 | 843 | 76.4 | -0.28 |
| 72.4 | 2265 | 2346 | 76.1 | 0.64 | 22000 | 72.8 | 1156 | 1190 | 74.4 | -0.36 |
| **71.4** | **1722** | **1757** | **72.9** | **0.55** | **23000** | **71.7** | **1607** | **1641** | **73.2** | **-0.45** |
| 70.5 | 1271 | 1306 | 72 | 0.45 | 24000 | 70.0 | 2138 | 2184 | 72.0 | -0.55 |
| 69.1 | 901 | 947 | 71.3 | 0.36 | 25000 | - | - | 4647 | 150 | -0.64 |
| 68.8 | 635 | 670 | 70.5 | 0.28 | 26000 | - | - | 5318 | 150 | -0.72 |
| 68.5 | 439 | 462 | 69.9 | 0.21 | 27000 | - | - | 6024 | 150 | -0.79 |

The breakeven bounds for the strategy example can be defined by subtracting the net premium from the put option strike and adding the net premium to the call option strike. **If** BTC/USD prices expand beyond the breakeven range, **then** the strategy would become in the money. Profits accumulate linearly, theoretically to infinity, with the long call option, and downside profit would be maximized if BTC/USD prices were to trade at $0.00. The payoff graph is shaped like a smile, as long straddle holders are happy when volatility expands.

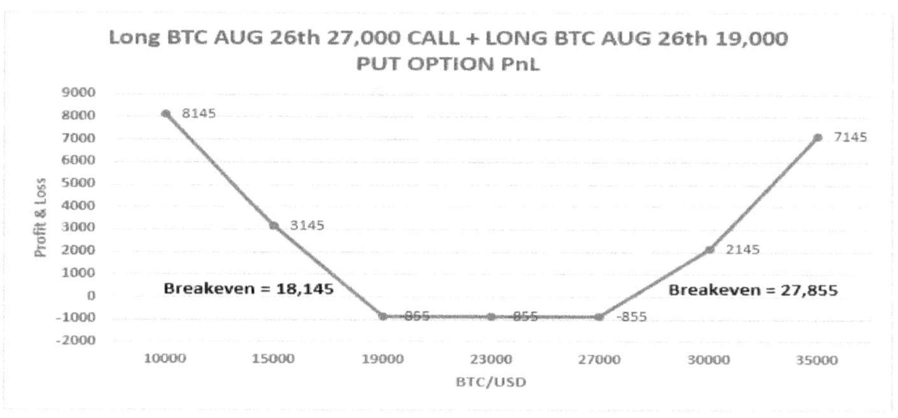

CHAPTER

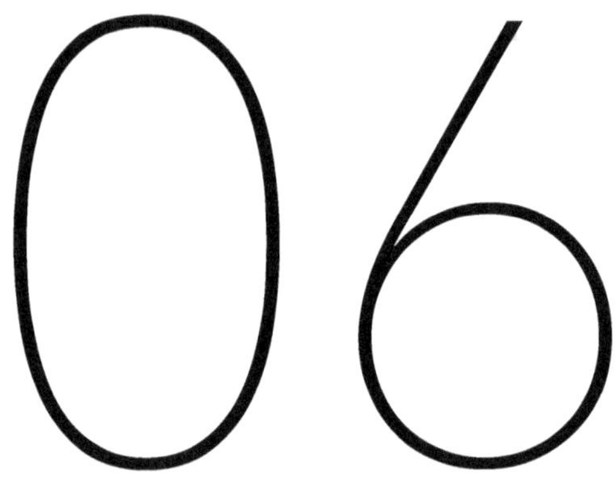

# FUNDAMENTAL MARKET FACTS

## HISTORICALS, CURVES & SKEWS

## FUNDAMENTAL MARKET FACTS

When engaging in strategy & execution analysis, market participants are typically familiar with time series charts for historical price action examination. Expanding on this approach, another set of tools that can be added are factual & verifiable *market internals*. Market internals are data points that seek out the heartbeat of a marketplace, providing insight into the sentiment of participants and the flow of transactions that can reprice the relative value of the underlying asset.

Market internals include raw data on derivative markets, such as open interest by contract and strike, volume by price and time range, historical and implied volatility, put-call skew, forward curves, upcoming expiration dates, and the associated open interest for those expiries. Market participants processing market internals can also reach for softer internals that explore seasonality and public interest or sentiment for an underlying asset.

Market participants seeking market internals information have two choices: publicly available and private paywall data. Typically, the best and most complete data will be found behind a paywall, as real costs are associated with aggregating and hosting such values. The source page at the end of this book provides resources for accessing publicly available data.

Regardless of one's approach to finding market facts, the total transactional data for an asset or marketplace is nearly impossible to aggregate completely as spot, options, futures, and structured product transactions don't trade on public venues alone. Dark liquidity pools serve a purpose to market participants by facilitating private and direct transactions in a peer-to-peer fashion. The nature of such direct transactions is rooted in privacy through bilateral execution between willing buyers and sellers, as opposed to execution via a public exchange. With this known unknown of data analysis, participants can better understand the marketplace by utilizing the following auxiliary data points to supplement their existing risk management strategy & execution order of operations.

# SEASONALITY

Before jumping into derivative market internals data, it is helpful to take a step back and broaden one's perspective on the asset under analysis by first understanding any seasonal effects that may impact prices from a fundamental angle. In stock markets, a seasonal effect from quarterly earnings reports impacts the relative value of company shares against a fiat currency. With commodities, the distance and position of the earth to the sun play a prominent role, as in the summer vs. winter seasonal effect on the supply and demand for grains, livestock, vegetables, fruits, timber, fibers, oil, natural gas, precious metals, industrial metals, rare-earth metals and so forth. In addition, natural and human-induced events that disrupt or shift the flow of goods and services will impact asset prices. The economy's ebb and flow at large and the prevailing interest rates for lending & borrowing also play a role in valuations for all assets as a reflection of growth (expansion) vs. stagnation (recession).

Regional and global economies are gauged via gross domestic product reports, producer prices, consumer prices, real estate values, interest rates, and unemployment levels. All these metrics influence the flow of capital; thus, risk managers' awareness of these data points can raise their perspective when making portfolio allocation decisions. For bitcoin, a known event called "the halving" occurs every four years and decreases the supply of bitcoin until the fixed quantity of 21 million bitcoin is reached. Alert market participants anticipate this event since it directly impacts risk management opportunities through derivative markets.

Tax seasonality may also impact bitcoin, as the asset has attracted a sizeable amount of speculators, given the wide price swings and volatility. These speculators and short-term traders are subjected to elevated levels of taxation relative to long-term holders. This potential for selling pressure to meet taxation requirements has also led many astute participants to leave open the possibility for income tax seasonality to influence BTC/USD prices.

# GOOGLE TRENDS

Another characteristic of bitcoin that has been highlighted is human psychology's susceptibility to being enamored with high volatility and the seemingly inevitable fear of missing out associated with such wide price ranges. Given this marketplace FOMO, staying on top of general interest from internet traffic in legacy search, engagement platforms, and media venues is also a soft data point for fundamental facts.

Trends.Google.com offers an exciting analytics database for active searches by keyword going back to 2004. Satoshi Nakamoto mined the Genesis Block of the Bitcoin Blockchain on January 3rd, 2009. This Genesis Block mined, and every subsequent block during the first four years of Bitcoin's existence yielded a block reward of 50 bitcoin every ten minutes. From there, the Block reward distribution was halved every four years.

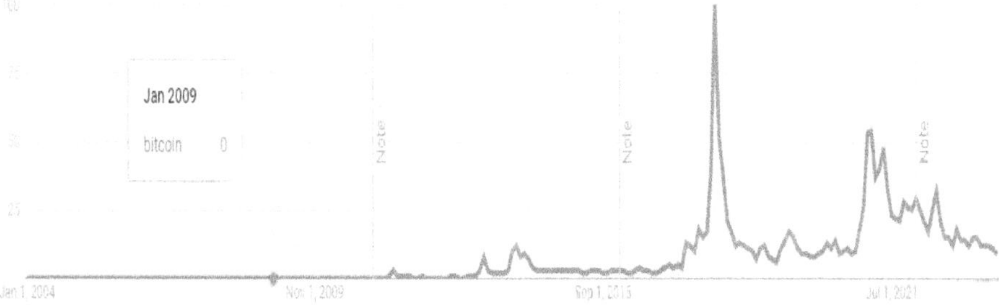


The digital asset bitcoin has existed in the public sphere for 14 years and counting. Google's trends for bitcoin help to understand the broad public interest. This general curiosity and search for information around bitcoin still seems motivated primarily around price action and the subsequent press coverage that garners. As more people understand the concept of money and learn about the monetary system under which they live, time shall tell how the Google trends for "bitcoin" will play out.

For market analysis, Google Trends can gauge broad public interest in assets and relative public interest in the present moment compared to the past or to other assets. Based on human psychology and the ever-present fear of missing out, public search data from Google aligns with bitcoin bull markets and dissipates as prices move lower or stagnate. Think back to 2021, where a common theme amongst traders was questioning valuations for bitcoin and other digital assets relative to public interest and impact. Google trends can help support or undermine a public interest thesis. Below is the Google trends data for the keyword: "inflation."

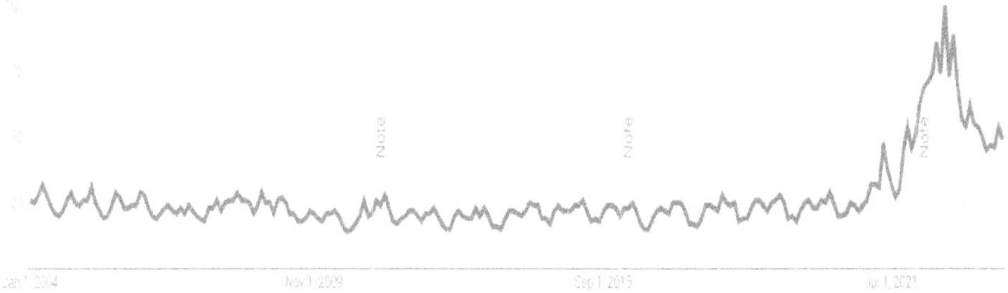

## BITCOIN MARKET CAPITALIZATION & USD PRICE

When analyzing corporate stocks, total market capitalization can be calculated by multiplying the total outstanding shares by the current market price. Market capitalization provides insight into the current cost of buying the company outright or taking a majority stake at the present price per share. A company with 10 million outstanding shares currently trading for $10.00 per share would have a market capitalization of $100,000,000.

As discussed in the introduction to this book, bitcoin market capitalization is easy to calculate as the protocol ruleset defines the total quantity of bitcoin as 21 million. Thus, for net market capitalization calculation, we multiply 21 million by current BTC/USD reference prices. Market capitalization is a helpful market fact of the underlying asset and can easily be forgotten when the BTC/USD reference price is in hyperfocus.

Understanding relative market capitalization can also help market participants determine the relative value of an asset, not just to the most common denominator but also to similar investments. For bitcoin, it is interesting to consider market capitalization relative to gold (13T), silver (1T), NVDA (2.4T), AAPL (2.72T), and GOOGL (1.71T).

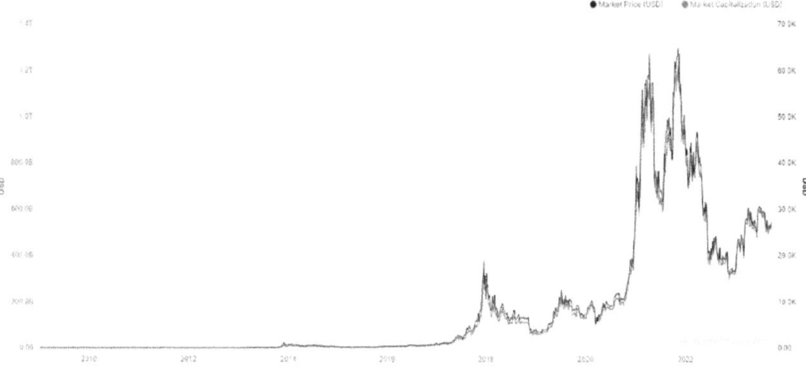

## ALT MARKETS

Below is a time series chart depicting the total digital asset market capitalization dating back to 2013, which reached a high of nearly $3T on November 9th, 2021.

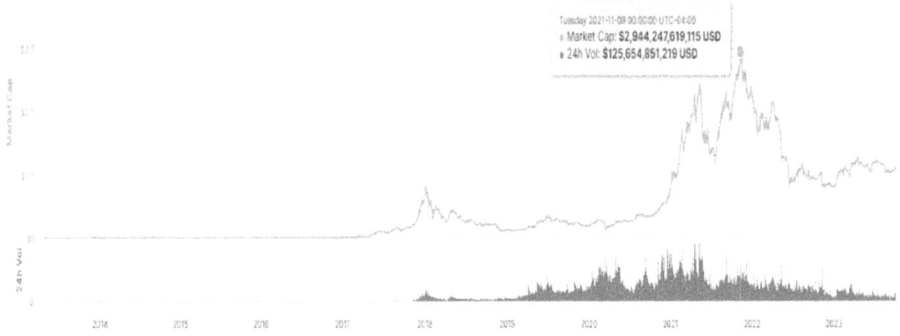


Another notable market fact relating to market cap is the total market capitalization of digital assets minus bitcoin. By removing the dominant asset, we can isolate the alternative asset market and define the total market capitalization of that group. Alt markets reached an all-time high on November 11th, 2021, totaling $1.6T net market capitalization, heavily weighted to Ethereum's native asset, ether - ETH/USD.

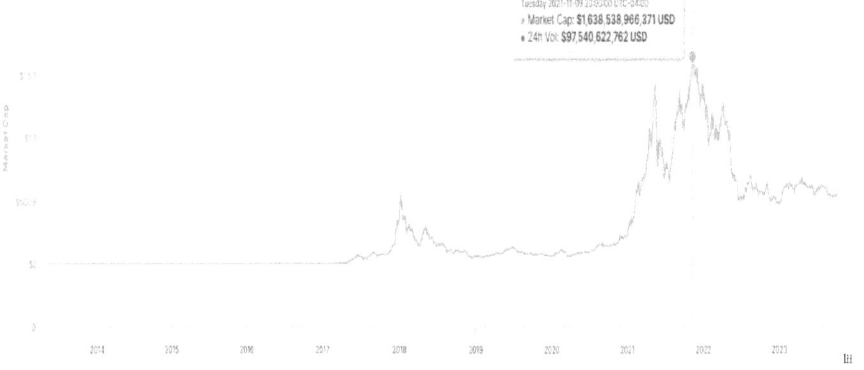


lii

# BITCOIN DOMINANCE & VOLUME

Expanding further on the concept of relative market capitalization, the chart below shows the individual proportions of the top ten digital assets by market cap relative to the total market capitalization of all digital assets that are actively traded and tracked by coinmarketcap.com. Market participants can use bitcoin dominance to assess market sentiment further, identify relative value opportunities, and validate or invalidate any thesis about capital flows.

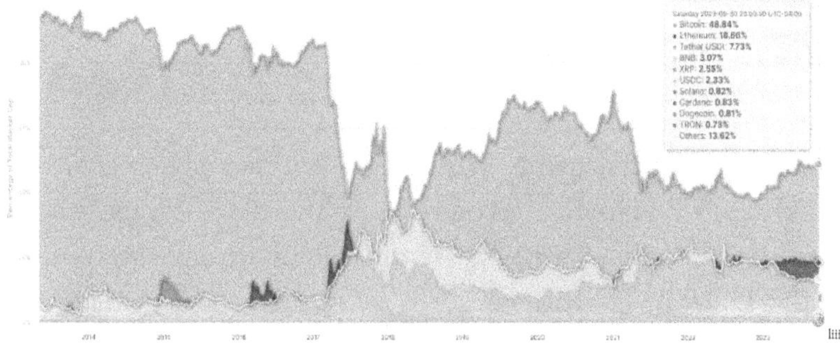

[liii]

Aside from dominance, the total notional value of bitcoin volume can also be factored in as a market analysis component when gauging for participation. Bitcoin volumes made all-time highs in February 2021, with data showing just north of $80B. This is likely an underestimate given the transactions that occur with no public transparency from bilateral execution.

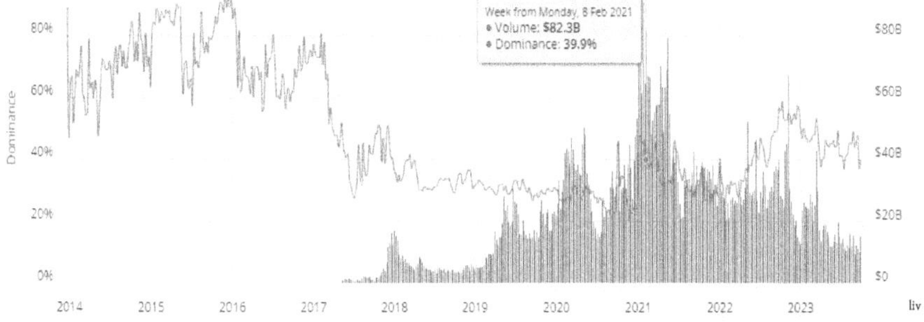

[liv]

# CORRELATION

Correlation can help identify the price movements of assets in relation to each other. This approach to relative value can assist in defining strong, weak, positive, and negative relationships across same-sector and multi-sector assets.

| Correl. Window: 30 | | | | | | | | |
|---|---|---|---|---|---|---|---|---|
| | | | 0 | | | | | |
| | BTC | ETH | SOL | Nasdaq | Tesla | SnP | Gold | USDT10y |
| BTC | 100% | | | | | | | |
| ETH | 89% | 100% | | | | | | |
| SOL | 78% | 80% | 100% | | | | | |
| Nasdaq | 22% | 16% | 27% | 100% | | | | |
| Tesla | 14% | 3% | 17% | 65% | 100% | | | |
| SnP | 33% | 27% | 32% | 95% | 57% | 100% | | |
| Gold | 26% | 22% | 28% | 28% | 19% | 39% | 100% | |
| USDT10y | -40% | -51% | -39% | -26% | -7% | -34% | -56% | 100% |

lv

With this data, market participants may seek to understand the attributes between BTC and ETH, the second-largest digital asset by market capitalization at the time of writing. With such a strong and positive correlation, the ETH/BTC ratio spread is an attractive risk management reference rate and tradeable instrument.

lvi

# FORWARD CURVE

Our exploration into options markets uncovered the importance of futures markets and the relationship that forward prices for an underlying asset have on options prices with matching expiration dates. With the concept of "no arbitrage" when pricing options, the forward curve is a critical data set when analyzing derivatives markets.

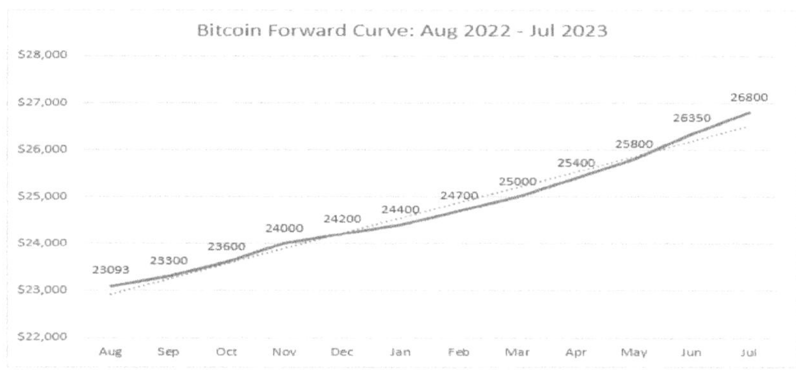


Forward curves can also be viewed from an annualized yield perspective. Since bitcoin holders can lock in forward sale prices relative to the current reference rate, the curve reflects the annualized risk-free rate of return per expiry.

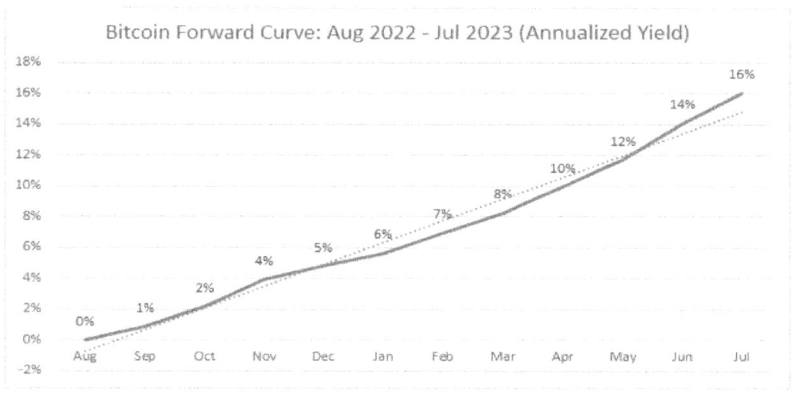

# OPTIONS VOLUME

Total daily option volume is a helpful metric for market participants seeking to understand market liquidity. Starting from a high level, analysts can aggregate daily exchange volume for bitcoin options and plot the data on a time series chart from a random sample, as shown below.

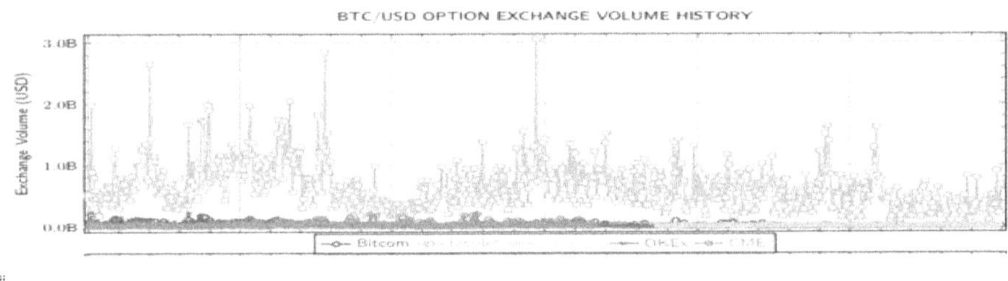


lvii

Options data can be isolated by strike and expiration date for a more granular view of volume, helping identify strike prices with the most trading activity and liquidity. Acknowledging these price levels and options expiration dates adds context to volume by adding strike prices that could attract or repel the BTC/USD reference price.

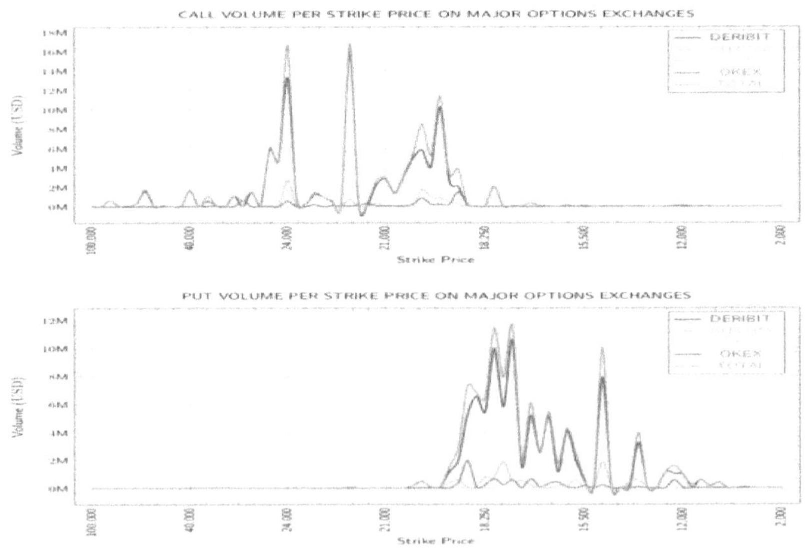

# OPTIONS OPEN INTEREST

While volume calculates the total number of contracts traded over a given time, open interest calculates options contracts held as open positions from the prior day. Where volume can alert to increased or decreased market activity, open interest can be more precise in identifying real liquidity zones. For these same reasons, futures market volume and open interest are other market internals to keep in mind for the full scope of derivatives liquidity.

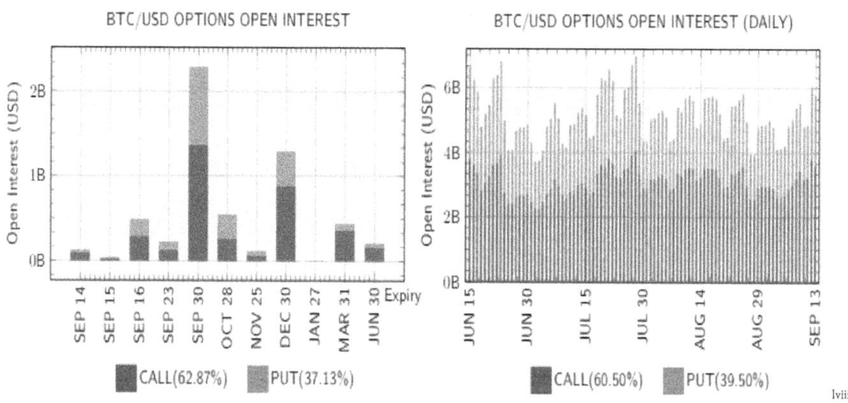

lviii

The charts above depict two distinct visualizations of open interest for calls and puts at varying expiration dates. Open interest for calls and puts at specific expiration dates is shown on the left side. To the right is a historical aggregation of daily open interest for calls and puts from June 15th to September 15th, 2021.

Aggregated open interest can assist in grasping intraday liquidity and activity. In contrast, open interest by expiration date provides market participants with information on the total notional value of options at expiration. This insight is significant as the nearest expiration date approaches. Open interest by expiration data typically shows larger net notional values for quarterly and yearly options, followed by monthly, weekly, and daily options.

For open interest by expiration, data can be narrowed further into open interest by strike at the defined expiration date. Analysts can zoom in on liquidity zones, market maker rebalancing zones, and net notional value of options by strike through the visualization of open interest by the strike. An example of this dataset can be viewed below for bitcoin call & put options that expire on February 26th, 2021.

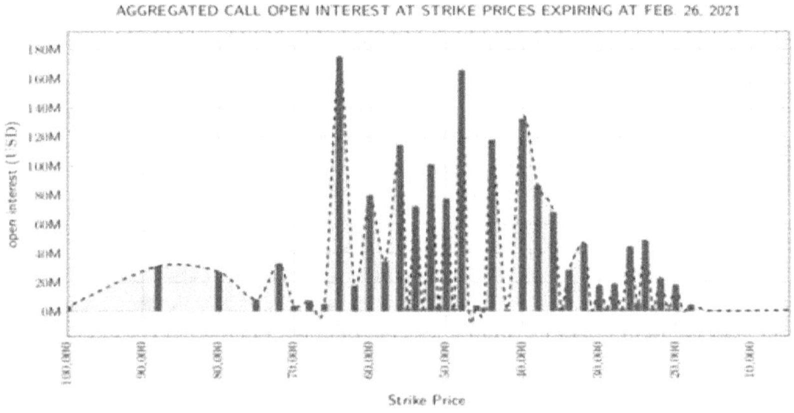

[lix]

As with all internals, and open interest in particular, the quantity of options traded via bilateral execution is nearly impossible to calculate and model. Although public data is widely available, analysts must be aware that public data still needs to be completed.

## HISTORICAL VOLATILITY

Historical volatility is an excellent metric to compare against current implied volatility. It is calculated from actual past price movements, most commonly over the prior 30 days. Once a historical time range is defined, traders can then annualize that value, as shown in the chart below.

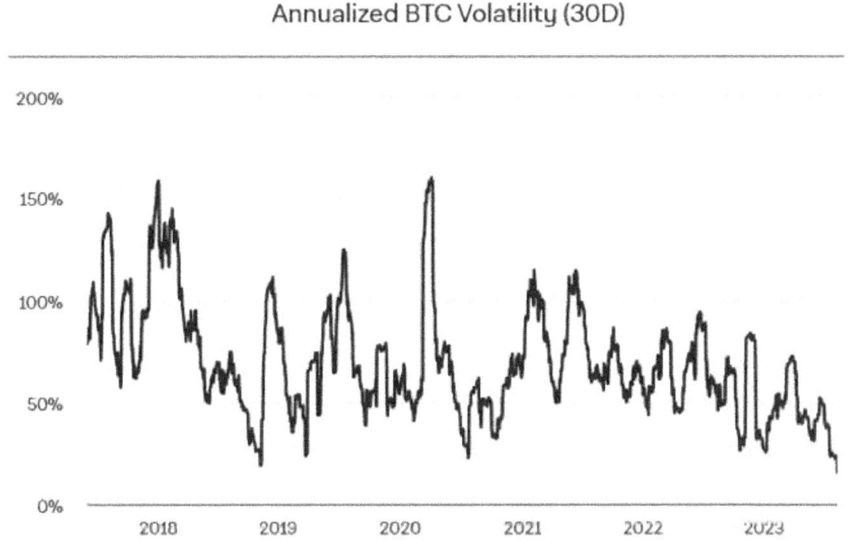


Along with the reference price movements for BTC/USD, the volatility of bitcoin is also quite volatile. Since 2017, bitcoin's volatility has ranged from highs above 150% to lows below 40%. This perspective provides value to market participants seeking to explore risk management strategies via options markets, relative to the volatility of bitcoin.

## IMPLIED VOLATILITY

Many active derivatives market participants will begin each day by analyzing implied volatility across expiration dates. ATM options typically hold the most extrinsic value compared to OTM or ITM options. Understanding the market's volatility expectation, or expected annualized standard deviation of price returns, is imperative for passive and aggressive execution techniques that deploy options strategies across varying strikes and expiries.

Building on volatility analysis, participants are better equipped to explore relative value opportunities by understanding the market's expectation for volatility across ATM options of gradually increasing time to expiration. The term structure, or forward volatility curve, can give deeper insight into overall market sentiment by plotting the data points for ATM volatility onto a graph.

Implied volatility experiences contango and backwardation as a forward curve. Near-term options can price volatility over further-dated expirations during great uncertainty, causing a backwardation structure.

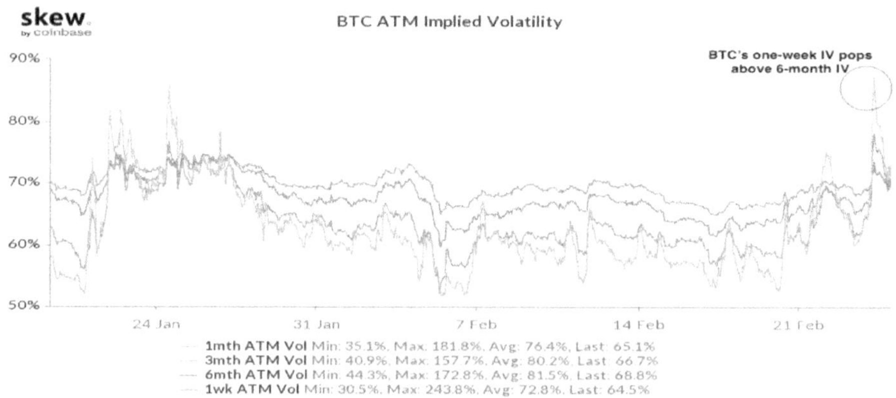


During times of more confidence in *known knowns* & *known unknowns* around an asset, further expiration volatility will trade at a premium to near-term options, given the wider expected range with more time to expiration.

At the time of writing, implied volatility is trading at elevated levels for Q2 2024 options given the *known known* of the 4th bitcoin halving event and the wider expected range for BTC/USD anticipated from the structural supply shock.

The chart below visualizes data on ATM implied volatility for 7-day, 30, 60, 90, and half-year expiry options, plotted with a historical perspective from 2019 to early 2023.

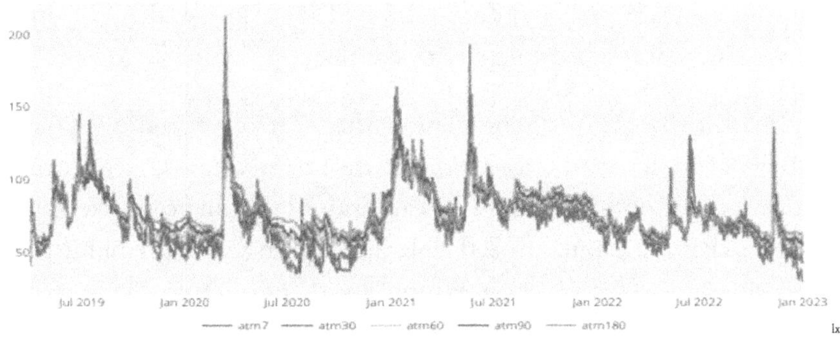


[lx]

For out-of-the-money volatility, fixing a strike price for both calls and puts allows data to be plotted for IV at each strike price across a range of expiration dates, as shown below. Fixing a Delta is also an interesting way to interpret implied volatility.

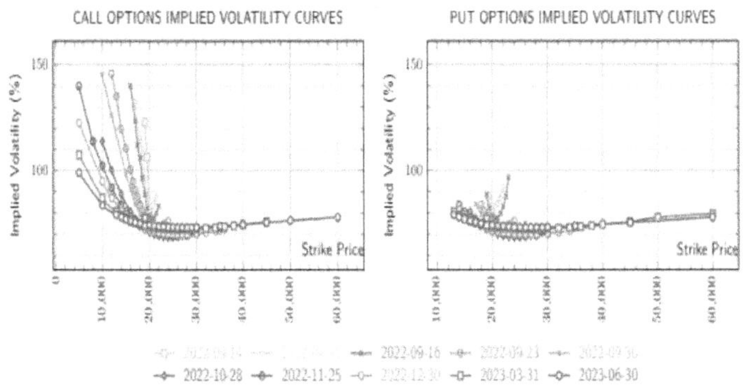


[lxi]

## SKEW

Skew is such a hot topic in options markets that the first major derivatives data provider for digital assets was named skew.com. This company eventually became a major Coinbase acquisition on April 30th, 2021. From an options data perspective, skew measures the relative value between calls and puts with the same expiration date at varying strike prices. The price difference of fixed strike options or implied volatility difference can measure this relative value. Supply and demand for ITM, OTM, and ATM options drive variability for implied volatility by contract.

When market participant sentiment shifts bearish, expecting a greater probability for downward price action for the underlying asset reference rate, the price for put options relative to call options will increase, skewing the options market with put implied volatility trading at a premium to call implied volatility. This can be measured by fixing a Delta to compare IV based on the moneyness of puts and calls. For example, skew can be measured for OTM calls and puts, such as the 25 Delta options. Similarly, skew can be calculated for ITM and OTM calls and puts with near or identical absolute Deltas.

For bitcoin, positive skew became highly pronounced in the bull run of 2021. Call options traded with significantly higher implied volatility than put options with the same Delta. When skew is *positive*, OTM call options trade at a higher IV than OTM put options. OTM puts trade at a premium to OTM calls when the skew is negative. Positive skew displays bullish sentiment bias, while negative skew displays bearish sentiment bias.

IV can be plotted for OTM call and put options at varying strike prices to create a volatility surface. The surface resembles a smiling face if OTM options trade with a higher IV than ATM options. If OTM option IV traded flat to ATM IV, the surface would resemble a "smirk."

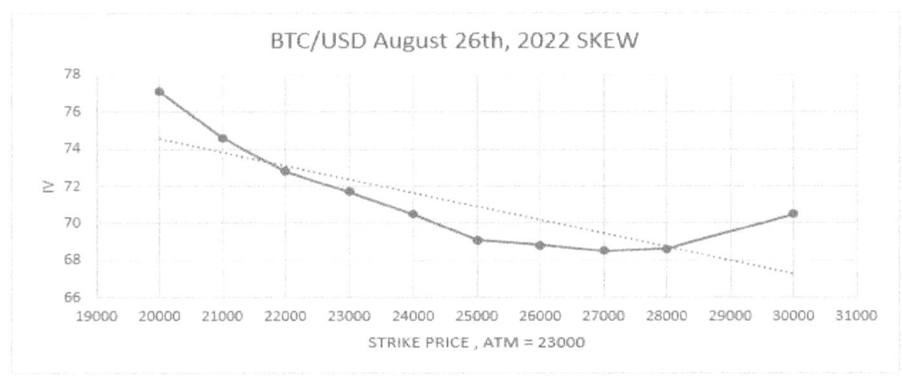

The chart below from Amberdata clearly shows movements in skew from year to year, beginning in 2019. Regarding volume and open interest, 2019 was still considered very early days for bitcoin options, with few participants executing at size and muted volatility throughout the year. In 2020, with the COVID event and subsequent interest in digital assets from stay-at-home restrictions implemented by governments worldwide, bitcoin options volumes moved higher, and more sophisticated participation led to broader movements in skew, albeit range-bound.

2021 was an explosive year for bitcoin, making new high after new high, as call options were feverishly bid over puts, causing positive skew with call implied volatility higher than put IV.

2022 continued to be volatile both in the underlying and skew, although to the downside. Demand for protection skyrocketed on headline news of imploding protocols, exchanges, hedge funds, and trading groups following multiple systemic failures and widespread counterparty risk contagion across the marketplace. Negative skew dominated market structure throughout the year.

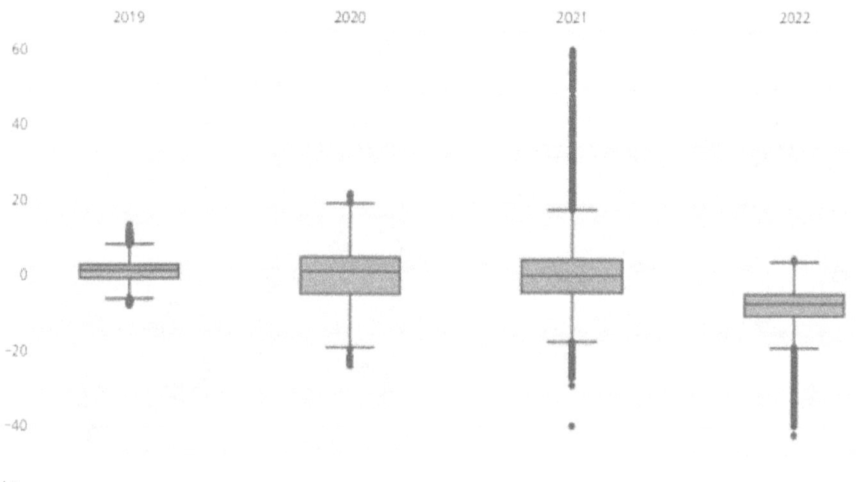


lxii

## SKEW INSIGHTS

Analyzing the difference in implied volatility between call and put options with the same expiration date, skew can be further understood by considering the fourth bitcoin halving event on April 20th, 2024. The table below shows that the 25 Delta skew is negative for near-to-midterm expirations, meaning 25 Delta puts have greater implied volatility than 25 Delta calls. Skew then turns positive for further dated options past September 2023 expiration.

| 25 DELTA SKEW | |
|---|---|
| EXPIRATION DATE | SKEW |
| 19-Aug-23 | -5.92 |
| 20-Aug-23 | -4.93 |
| 21-Aug-23 | -4.93 |
| 25-Aug-23 | -5.06 |
| 1-Sep-23 | -3.40 |
| 8-Sep-23 | -1.72 |
| 29-Sep-23 | 0.12 |
| 27-Oct-23 | 1.75 |
| 29-Dec-23 | 3.05 |
| 29-Mar-24 | 4.03 |

The table above shows a symmetrical skew for the September 29th, 2023, options expiration. This near-zero value shows that 25 Delta puts and calls have equal or near-equal implied volatility. From October 27th, 2023, the expiration date forward through the dataset skew turns positive with call implied volatility trading at a premium to puts. This information from the 25 Delta skew curve can be used to support the thesis that the market is anticipating more significant demand for upside optionality ahead of the Bitcoin having event relative to near-dated options.

Based on skew analysis, active participants can measure skew movements daily for fixed strike options. Skew movements calculate real-time supply/demand shifts for puts and calls with the same Delta and expiration date. These shifts are demonstrated below via an implied volatility move from the prior close for ATM options per each corresponding expiration date. This dataset is beneficial in gauging shifts in skew or continuations in skew structure throughout the forward curve.

For 2-Oct-23 & 3-Oct-23 ATM options, the curve data displays call volatility moving higher by 10.5 and 9.04 percentage points, respectively, from the prior day's close. From there, call volatility traded higher relative to put vol for the remainder of October 2023 expiration dates. Skew continued to narrow down the curve until turning negative for Jun & Sep 24 expiries – meaning that from the prior-day close, put option volatility moved higher relative to call volatility by the respective data value.

| FIXED ATM VOL MOVE | |
|---|---|
| EXPIRATION DATE | SKEW |
| 2-Oct-23 | 10.5 |
| 3-Oct-23 | 9.04 |
| 6-Oct-23 | 5.33 |
| 20-Oct-23 | 1.11 |
| 27-Oct-23 | 1.01 |
| 24-Nov-23 | 0.69 |
| 29-Dec-23 | 0.27 |
| 29-Mar-24 | 0.24 |
| 28-Jun-24 | -0.11 |
| 27-Sep-24 | -0.27 |

These shifts can then be plotted on a time series chart per expiration date:

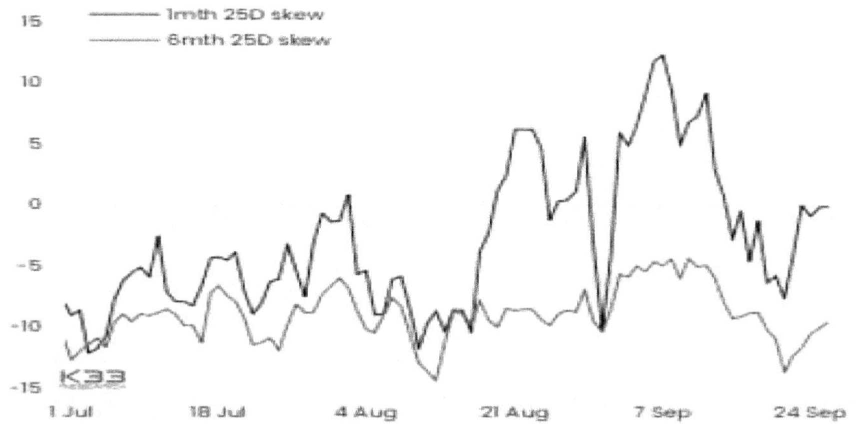


lxiii

## RISK REVERSALS

Merging the concept of skew with implied volatility, ATM volatility term structure across the forward curve of expirations acts as the spine of the volatility surface structure. Stemming from the spine are the wings of volatility, the OTM call-and-put options. At the strategy development level, selling OTM calls and buying OTM puts for zero cost is the **OPTIMIZED HODL** strategy, which is a short risk reversal.

Applying skew analysis, the **OPTIMIZED HODL** can be executed into favorable skew structures. For long bitcoin positions, selling calls and buying puts provide upside profit potential to the call strike and downside protection at the put strike. Execution can be optimized when selling calls and buying puts into positive skew, where calls are more expensive than puts.

Conversely, the **OPTIMIZED FOMO** strategy is long a risk reversal, where a market participant *sells* and OTM put while simultaneously buying an OTM call with nearly identical Deltas. When the premium of the sold option is greater than the premium required to purchase the second leg of the structure, that can be thought of as a zero-cost collar.

Source: Laevitas

The yearly skew analysis above helps visualize the impact of relative call and put implied volatility when assessing risk management opportunities. Whether selling the risk reversal, as with the Optimized HODL, or buying the Risk Reversal, as with the Optimized FOMO, the market's skew will directly impact the edge of spot moves that make the position profitable or unprofitable.

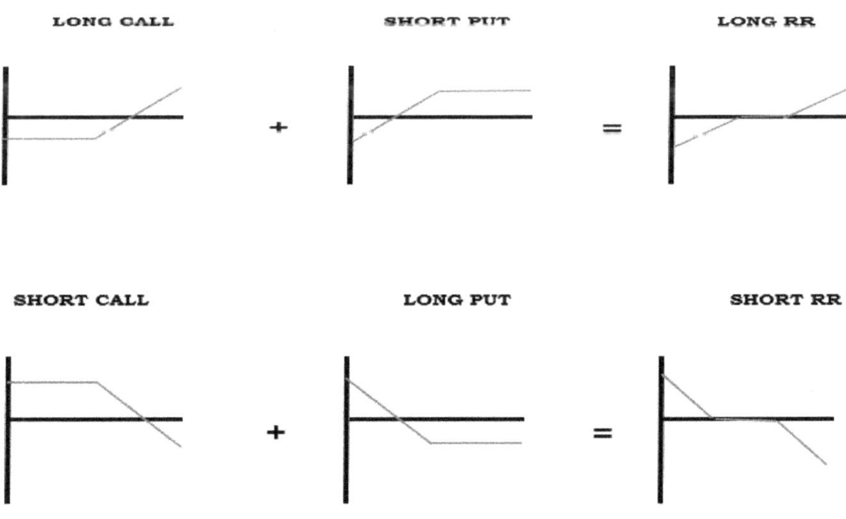

Energy Storage, Risk Management, Real Yield.

## OPTIONS ACTION

Prior day options action, meaning analysis of unusual volume and open interest shifts by strike & expiration date, can be used to help determine where "smart money" is positioned. The more notional volume and open interest held per contract, the more significant the market impact is due to portfolio rebalancing from liquidity providers and general supply/demand imbalances seeking price equilibrium. When trying to follow "smart money," analysts can observe intraday transactional data and seek to identify outsized executions of contracts. These data points provide insight into the long or short-positioning of well-informed, well-capitalized market participants.

# CME COMMITMENT OF TRADERS

Another tool to reveal market participant positioning is the CME COT report, which tracks long and short positioning for various participants, such as asset managers, dealers, leveraged accounts, and reportable entities.

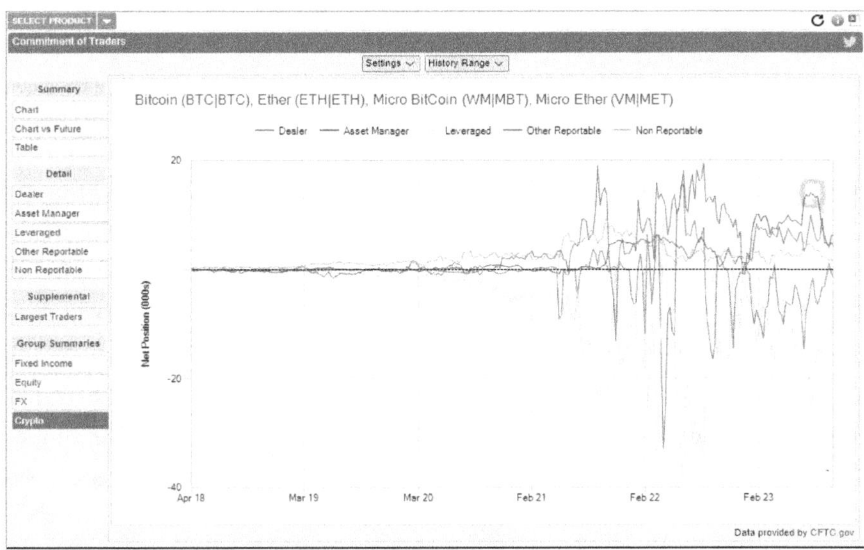


*"The Commitments of Traders (COT) tool provides a comprehensive and highly configurable graphical representation of the CFTC's report on market open interest released each Friday afternoon based on open positions as of the preceding Tuesday. Markets are only included if 20 or more traders hold positions equal to or above the reporting levels established by the CFTC and the respective exchanges.*[lxiv]*"*

# DENOMINATOR ANALYSIS

Since the relative value of BTC/USD is of utmost importance to bitcoin holders who are seeking to manage purchasing power risk with bitcoin as a store of value asset relative to the leading world reserve fiat currency asset, it is crucial to focus on the numerator, bitcoin, and the *denominator* - U.S. dollars. Keeping track of the open market value of USD relative to other major global currencies can provide more granularity into near-term asset valuations where USD is the denominator.

The charts below are for the U.S. Dollar Index, symbol DXY, a weighted index that measures U.S. dollar valuations relative to fiat currencies of major United States trading partners. As relativity is a core theme when speaking of valuation, the appreciation in price for dollars or the depreciation of dollars relative to other fiat currencies can be visualized in the second chart. When U.S. dollars are strong, as the denominator against broad asset valuations, U.S. dollar strength can be a headwind for asset prices, such as oil, bitcoin, or gold. A weakening dollar can be a tailwind for asset price valuations, all else equal, as the value of the denominator decreases.

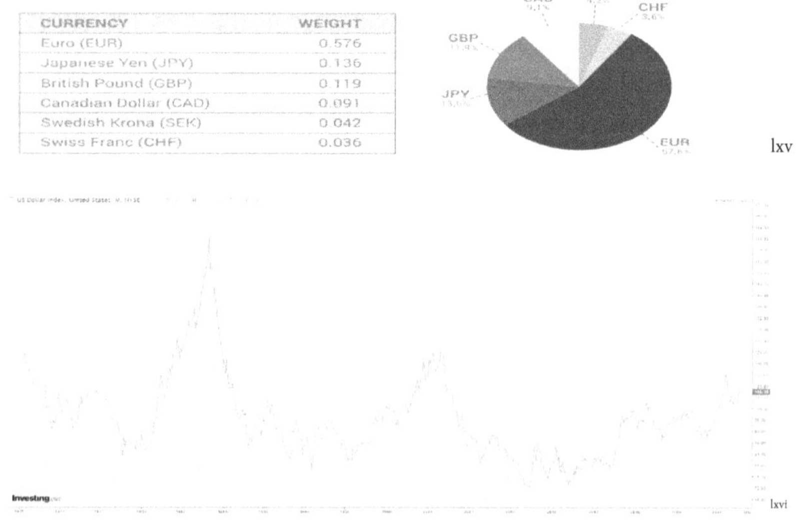

| CURRENCY | WEIGHT |
|---|---|
| Euro (EUR) | 0.576 |
| Japanese Yen (JPY) | 0.136 |
| British Pound (GBP) | 0.119 |
| Canadian Dollar (CAD) | 0.091 |
| Swedish Krona (SEK) | 0.042 |
| Swiss Franc (CHF) | 0.036 |



lxv

lxvi

Central bank control over interest rates plays a significant role in any given fiat currency's relative value to other fiat currencies, as interest rates are the cost & yield of borrowing or lending capital. This force on capital flows directly influences the supply/demand for the underlying currency asset itself. When interest rates are elevated, the currency supply is constricted by waning demand associated with elevated borrowing costs. Yield opportunities from capital lending cause risk managers to reassess capital allocation to risk assets compared to lending the denominator asset for interest.

With the DXY so heavily weighted to the Euro, by comparing interest rates for 10-year maturity U.S. Treasury bonds and 10-year German bonds, we can identify that interest rates for U.S. dollars have remained elevated above interest rates for European fiat, unveiling a tighter supply and greater yield in risk/reward analysis. The interest rate differential causes shifts in supply/demand between each currency. When the interest rate differential for U.S. dollar-denominated bonds exceeds that of Euro-denominated bonds, expectations would be for U.S. dollar values to trend higher relative to Euros and vice versa. This logic derives from the supply/demand to lend and borrow the currency asset and helps understand U.S. dollar index price movements.

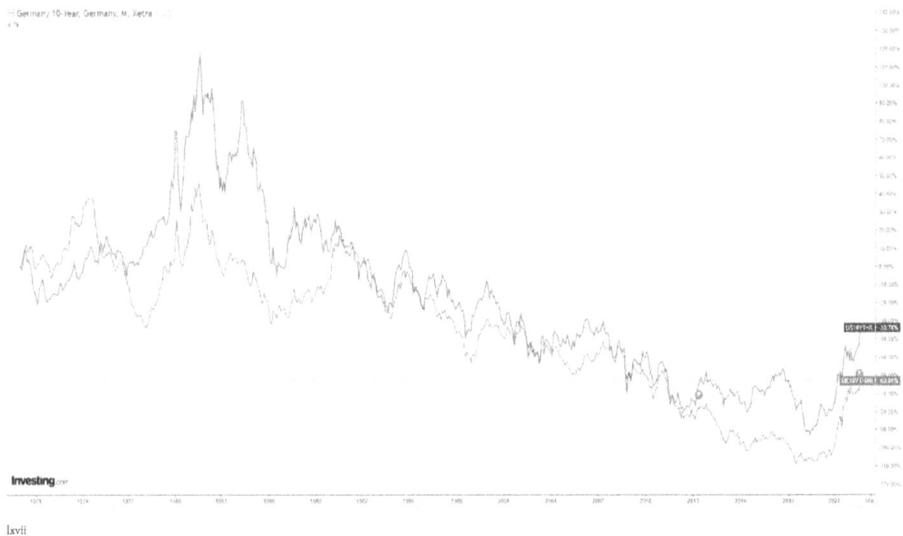


lxvii

To finalize an overview of fundamental market facts culminating with a focus on the denominator asset for the BTC/USD reference rate, the force that drives the relative value of the U.S. dollar is examined. By zeroing in on the relationship between interest rates for the U.S. dollar and the U.S. dollar index price itself, the goal is to grasp how monetary policy has the potential to dramatically shift asset valuations that are measured against that denominator currency.

The following chart compares the annualized percent yield movements of 10-year U.S. Treasury bonds to the dollar index value (DXY.) Notice the volatility of rates far exceeds the volatility of the DXY index. However, the relationship is also apparent despite the volatility difference. Historical data shows that rising U.S. treasury bond interest rates lead to rising DXY values, and falling bond interest rate yields lead to lower DXY values. The following chart amplifies the core reason why bitcoin exists today: central banks control interest rates, thus maintaining the value of fiat currency by manipulating supply and demand for the currency itself.

If the store of value thesis for bitcoin rests on U.S. dollar devaluation through inflation, then this chart can be a window into the scenario. Since deficit spending comes with interest payments, monetary policy decision-makers face a critical juncture. Should central banks influence bond markets to keep interest rates low, alleviating the interest payment burdens associated with debt accumulation? The cost of that decision is the threat of inflation or hyperinflation. Alternatively, central banks may constrict the flow of money by selling government bonds they hold on their balance sheet, which accelerates increasing interest rates across a global population buried in national, local, business, and personal debt obligations.

CHAPTER

# EXOTIC OPTIONALITY

## PATH (IN)DEPENDENCE

To add some spice to the life of a risk manager, discovering the exotic side of optionality can broaden the application for strategy development when executing the accumulation, sale, hedge, or yield of a portfolio asset. To better understand exotic options, the critical contrast to vanilla options must be understood: reference price path dependency vs. path independent payout during the life of an option.

European-style *vanilla* call & put options are path-independent options. The payout of a vanilla call or put option is calculated from the difference between the strike price and the settlement reference rate of the asset at expiration. It does not matter what price path the asset took through the life of a European-style vanilla option; it is only the difference between strike and reference price at expiration that determines the payout of the option.

*Exotic* options are path-dependent, meaning the option's payout potential varies based on the reference rate movements of the underlying asset during the life of the option.

**LOOKBACK OPTIONS**: The exercise price of the call or put option is equal to the lowest or highest reference rate through the option's life. This is a sell high, buy low option, where a LOOKBACK call option owner has the right to purchase the underlying asset at the lowest reference price point during the option's existence. Similarly, and inversely, a LOOKBACK put option owner has the right to sell the underlying asset at the highest reference price recorded during the option's existence. LOOKBACK options "look back" in time and provide optionality at optimal price points over customizable periods from the option's life. These options demand higher premiums than vanilla call or put options for this exotic optionality.

**AVERAGE RATE OPTIONS**: Also referred to as *Asian style* options, average rate options are just as they seem; the average reference rate of the underlying asset during the option's existence determines the strike price of the Asian call or put option at expiration. The strike price of the Average Rate option at expiration can be customized to calculate the average of the reference rate over the entirety of the option's life or a specified time-period during the option's existence.

Suppose a bitcoin HODL'r wanted to ensure a BTC/USD conversion price for a certain percentage of bitcoin holdings. The HODL'r could execute a one-year average rate put option, obtaining the right to convert BTC/USD at the average reference rate over the following 365 days. A HODL'r seeking to buy more bitcoin could purchase a one-year average rate call option, granting the right to purchase bitcoin at the average price over the following 365 days.

## DIGITAL OPTIONS

Digital Options are customizable strike and expiration binary options with varying premium costs based on those customizations. Binary options provide a true/false payout fixed as 0% or 100% of the notional contract value. The payoff is triggered true for digital call and put options when the reference rate is above the strike price for digital call options and below the strike price for digital put options. As with most vanilla options contracts, most digital options positions can be closed ahead of expiration at some rate of probability-based value of a payout, provided there is ample liquidity for the contract. Digital options can be bought or sold at trade inception, assuming there is liquidity to enter long or short the contract at bid & offer prices.

When a digital option is European style with a payoff only at expiration, the digital option is referred to as a ***PAY-AT-IN*** option. If the digital option is American style with an active payoff trigger at the strike price from the contract's inception, it is referred to as a ***PAY-AT-HIT*** option. Call and put pay-at-hit options payoff triggers true *immediately* upon the underlying asset's referencing rate breaching the strike price during the option's existence. In contrast, pay-at-in call or put options only trigger payoff at expiration.

Assume that the spot reference rate for BTC/USD is $25,000, and a 30000-strike European-style digital call option that expires in 30 days was purchased for 10% of the notional contract value of 1 bitcoin. Suppose a volatility event drove the price of bitcoin 20% higher overnight, with the BTC/USD reference rate trading higher to $30,000. In this scenario, the owner of the digital call may seek to calculate the present value of their digital option to determine if they would like to sell to close their position ahead of expiration.

With a move higher in reference prices from $25,000 to $30,000, a European-style digital call option with a strike price of $30,000 would be considered

AT THE MONEY, implying a 50% probability of bitcoin being above or below $30,000 going forward. For this, the value of digital options can be measured as 50% of the one bitcoin payout at expiration, which is 0.5 bitcoin or $15,000. If the reference price exceeded the strike price at expiration, the European-style digital call holder would receive a payout of 1 bitcoin (100% of the contract's notional value).

From the seller's perspective, assume that the BTC/USD spot reference rate is $27,700. A European-style digital call option could have a strike price of $31,000 that expires in 30 days, available to be bought or sold for 10% of 1 bitcoin (0.1 bitcoin) or 10% of the BTC/USD reference rate of $27,700 ($2,700). At expiration, the seller of the contract is obligated to pay one bitcoin or $31,000 to the buyer of the contract were the BTC/USD reference rate to be above $31,000 on the expiration date 30 days from the contract inception date.

**If** the BTC/USD reference rate is below $31,000 at the time of expiration, **then** the seller of the contract would keep the initial premium paid from the buyer of the contract at the trade inception date as a premium cost (0.1 bitcoin or USD 2,700). From the seller's perspective, this digital call sale would be similar to a vanilla option-covered call sale. However, the payout function and risk/return profile are vastly different. Digital call options provide unique customization and fixed true/false payouts.

The same logic can be applied to a European-style digital put option with a true/false payout at a predefined expiration date. With a BTC/USD reference rate of $27,000 and a strike price of $24,500, a digital put option could be bought or sold for a predefined premium amount – potentially 10% of the contract's notional value that expires in 30 days. The buyer of this digital put would receive a one bitcoin payout were the BTC/USD reference rate to be below the strike price of $24,500 at expiration. If the settlement price of the asset were to be above the strike price, the buyer of a European-style digital put option would lose the premium paid to purchase the optionality. Thus, the seller of the option would keep the premium paid from the buyer at trade inception.

Plain and simple vanilla call and put options can be used to design risk management and real yield strategies suitable for most market participants. However, the depth of customization is limited by the standardization and reference price path *independence* of European-style vanilla options contracts. By scratching the surface of more customizable options contracts like LOOKBACKS, AVERAGE RATE, BARRIER, and DIGITAL options, the realm of structured products can be discovered with practical application.

Exploring exotic optionality allows for comprehending further risk management and real yield opportunities that can be deployed through portfolio diversification efforts. Customized structures can add another layer of controlled execution when accumulating an asset, selling an asset, ensuring a BTC/USD conversion rate, or seeking yield from the underlying asset.

## KNOCK-IN & KNOCK-OUT BARRIER OPTIONS

Barrier options are path-dependent option contracts with exotic characteristics that can be either American or European style. Although some similar traits are shared with vanilla call and put options, the critical point of barrier options is that optionality is either activated or deactivated upon breach (or non-breach) of the barrier price level(s).

Four common types of barrier options are as follows:

- **Down & Out**: Right to buy or sell bitcoin at the strike price if the reference rate does not touch the **lower** barrier price level through the option's existence.
- **Up & Out**: Right to buy or sell bitcoin at the strike price if the reference rate does not touch the **upper** barrier price level through the option's existence.
- **Down & In:** Right to buy or sell bitcoin at the strike price if the reference rate touches the **lower** barrier price during the option's existence.
- **Up & In**: Right to buy or sell bitcoin at the strike price if the reference rate touches the **upper** barrier price through the option's existence.

Down-and-out, along with up-and-out barrier options, are often referred to as
**KNOCK-OUT** options, whereas down-and-in and up-and-in options are referenced as **KNOCK-IN** options.

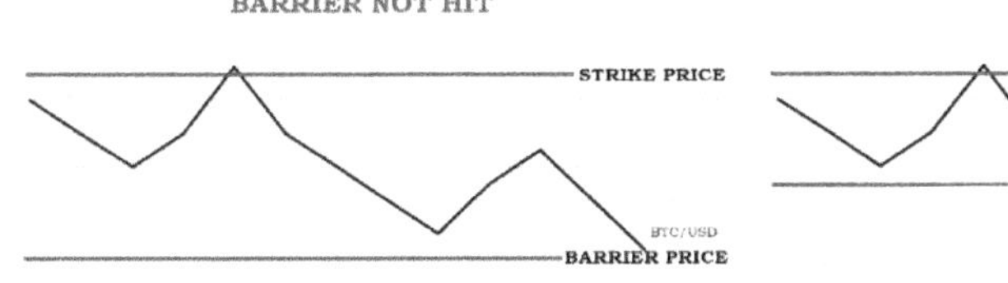


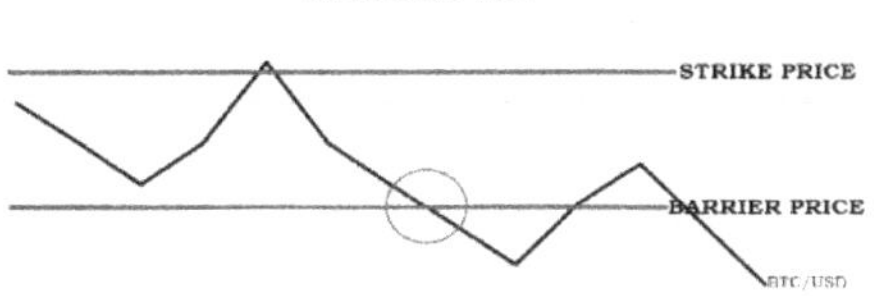


lxviii

**Knock-in**: BTC/USD Price ***has not breached*** the barrier price level; **optionality *is inactive*** at the strike price.

**Knock-out**: BTC/USD price **has not breached** the barrier price level; **optionality *is active*** at the strike price.

**Knock-in**: BTC/USD Price **has breached** the barrier price level; optionality ***is active*** at the strike price.

**Knock-out**: BTC/USD price **has breached** the barrier price level; optionality ***is not active*** at the strike price.

With unique payoff profiles and customizable barrier price levels that trigger the initiation, continuance, or deactivation of optional or obligatory events throughout the life of the trade, knock-in and knock-out options are commonly integrated into structured products for the accumulation or sale of an underlying asset. Knock-ins set a barrier price level from which call or put optionality is *activated*, while knock-outs define a barrier price level that *deactivates* and terminates optionality once breached.

From a practical perspective, risk managers may be interested in barrier options as similar optionality to vanilla options positions can be achieved at cheaper costs. With knock-in options, when the barrier price is far from the current reference rate, the option will trade at a significant discount to the corresponding vanilla option. If the knock-in barrier price is near the current reference rate, the knock-in option is likely to trade at a slight discount relative to the corresponding vanilla option. If the knock-in barrier price has been breached, the knock-in option will trade at the same value as the corresponding vanilla option.

Suppose an asset manager sought to buy upside optionality while the BTC/USD reference rate was trading at $25,000. The asset manager could purchase 30000-strike vanilla call options that expire in 30 days or a 30000-strike *up & in-call* option with the knock-in barrier price of $33,000. For this barrier option, the call optionality granting the right to buy BTC/USD at $30,000 is *not activated* until the knock-in barrier price level of $33,000 is breached. At the same time, a vanilla call option has an intrinsic value above $30,000. Due to this difference in optionality, the 30000-strike *up & in-call* option would be cheaper than a 30000-strike vanilla call option. However, were BTC/USD prices to trade above the knock-in barrier price level of $33,000, activating the 30000-strike up & call option, both the vanilla and barrier options would have equal value as the knock-in component grants equal optionality at the strike price.

For downside optionality, the same thought process rings true. A risk manager could buy vanilla put options or buy down-and-in barrier put options with a barrier price below the strike price. For example, if the BTC/USD reference rate was trading at $25,000, a risk manager could purchase a 22500-strike down-and-in barrier put option with a barrier price level of $20,000. This would be cheaper than buying the corresponding 22500-strike vanilla put option, while having the equivalent value of that vanilla put option were the BTC/USD reference rate to breach the lower barrier price of $20,000. The premium cost differential to buy the vanilla put vs. buying the barrier put option is due to the reference rate path dependence of the barrier option, meaning the optionality is not activated until the reference rate breaches the barrier price level.

A crucial customization to barrier options is the frequency from which a barrier breach is deemed triggerable. Barrier breach triggers can occur continuously from trade inception, meaning that the BTC/USD reference rate can trigger a barrier event at any time from trade inception. *Discreet* observation intervals can also be defined, where the barrier price level must be breached during a specific period during the option's life.

# ONE-TOUCH & NO-TOUCH BINARY BARRIERS

One-touch and No-Touch binary options use barrier price levels that knock in or out optionality. The options can be bought or sold and often combine a lower barrier price and an upper barrier price, which define a *price range* around the current reference rate of the underlying asset.

A One-Touch barrier option pays off if the underlying asset reference rate breaches the upper or lower bound barrier price levels. The breach event can be customized to trigger continuously from trade inception through the life of the option, during specific observation periods, or only triggerable at expiration. Barrier price levels that are further away from the underlying asset's price at trade inception have lower premium costs relative to barrier range levels closer to the underlying asset reference rate at the time of execution. Time to expiration also factors into the premium payment, with a more significant premium cost for continuous observation and longer-dated options relative to discrete observation periods and shorter-term expiration dates. A greater time premium is associated with continuous or discreet trigger observations relative to structures that only observe the reference rate positioning to the barrier levels at expiration, given the formers' increased probability for an actual payoff trigger event.

Inverse to one-touch optionality, *no-touch* optionality is a common range structure. Typically referred to as a **DOUBLE-NO-TOUCH,** this range structure provides a true/false payout event from barrier optionality defined by upper and lower bound price levels that provide a payout when the reference price *does not touch* either upper or lower bound. The trigger event can be customized through continuous reference price observation, during discrete observation periods, or at the expiration date. Were the reference price to not touch either the upper or lower bound barrier price levels in accordance with the structure's trigger observation schedule, then 100% of the notional contract value would be paid out to the DNT buyer.

By examining an example DNT structure offering where the current reference rate for BTC/USD is $27,700, and the upper & lower bound barrier price levels are set to $29,085 and $26,315, respectively, the expected range of the asset can be determined. Calculating the difference between the barrier levels and the current BTC/USD reference rate, market participants can utilize this structure as exposure to BTC/USD volatility. The expected range can then be measured against the premium cost to purchase (or sell) the optionality in exchange for the potential payout of the structure's notional value (or premium cost).

Market participants who seek **long volatility** as a hedge against spot bitcoin holdings can buy one-touch options and/or sell DNT structures.

Participants expecting muted volatility and want **short volatility** exposure can sell one-touch options and/or buy DNT structures.

## STRUCTURED PRODUCTS

Structured products are prepackaged trades combining different vanilla and exotic options to produce a payoff profile crafted for a market participant's specific needs. Other financial instruments such as treasury bonds, corporate bonds, and other yield-bearing assets can also be incorporated to the structure and payout of the structured product or used as an initial margin to open the position. This type of flexibility allows for the design of yield enhancement, zero-cost, and/or principal-protected coupon structures. In some cases, various assets (BTC, GOLD, OIL) could be used as collateral to margin any purchase or sale obligations assumed from the rights and obligations of a customized structured product.

Structured products began to take hold in legacy financial markets during the 1970s, with the tailwind in financial innovation following the renewed energy provided to the marketplace from the unveiling the Black-Scholes-Merton options pricing model. At first, structured products were created to embed optionality into treasury bonds. Other risk management use cases became evident throughout commodities and foreign exchange. By the 1980s & 1990s, structured products were designed around corporate stock holdings and liquid fund offerings.

Many types of structured product offerings exist with various components and complexity. For bitcoin, structured products are at the forefront of risk management & real yield opportunities, building from the sustained growth in vanilla options market liquidity. This section aims to scratch the surface of customizable trades designed around the BTC/USD reference rate. Four unique structures intended to be useful for accumulation, sale, or yield generation scenarios will be examined.

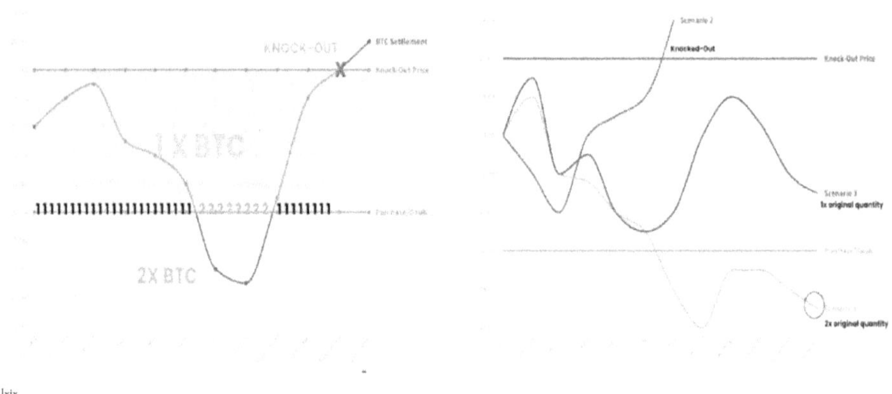


lxix

# ACCUMULATOR

An Accumulator is a structured product composed of exotic and vanilla options that allow for the purchase of bitcoin at prices below the current reference rate for BTC/USD. Against the current reference rate at trade inception, a **purchase/double-up** level is defined relative to the premium cost of the Accumulator structure (typically done at a purchase/double-up price level that allows for zero cost to buy the structure).

Accumulators are designed for market participants seeking to purchase bitcoin at a specific price over a predefined time below the current BTC/USD reference rate at trade inception. Accumulators can supplement or replace dollar cost-averaging strategies while tailoring duration and purchase price. The typical risk for an Accumulator structure, as is the case in the following example - **if** the BTC/USD reference rate settles below the *purchase price level* during any observation period through the life of the structure, **then** the Accumulator owner is obligated to purchase 2x (double up) the quantity at that price.

The theoretical payoff chart on the next page shows that an Accumulator has been designed to facilitate the purchase of bitcoin for $20,000, once per month for 12 months. At inception, the reference rate for BTC/USD was $23,000; thus, the purchase level is 13% below the current market price. The structure incorporates a *knock-out* barrier level at $25,000, which forces the structure to terminate and expire early if BTC/USD prices breach the knock-out barrier price level during any observation period.

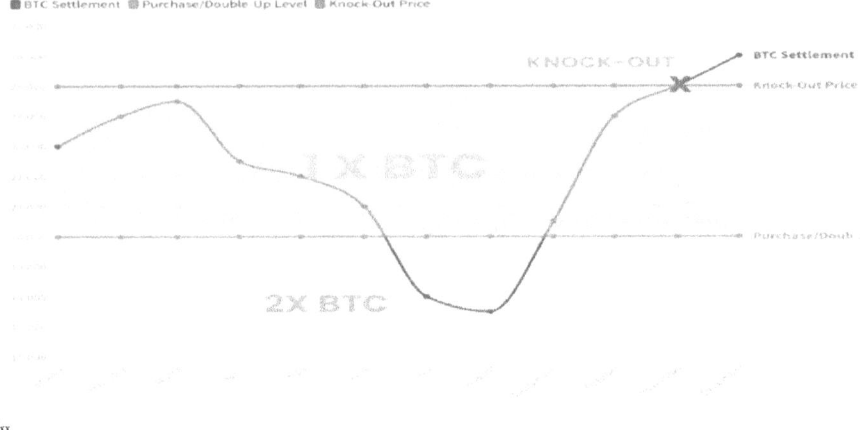


lxx

The *purchase/double up* barrier price level can be customizable, meaning the quantity of bitcoin that must be purchased at that barrier, if breached, could be 3x the initial amount, or 4x, etc. The Knock-out level could also be customized to *knock back in,* and the range between the knock-out barrier and purchase/double-up barrier can be widened or tightened, changing the premium cost for the structure.

Structured products are often sold as "zero-cost" structures (no premium cost) but typically require higher initial and variable margin as collateral for the position compared to vanilla options. A market participant could pay a premium cost to widen out the purchase/double up and knockout barrier levels. Conversely, a market participant may opt to tighten the knockout and purchase/double-up barriers levels relative to the BTC/USD reference price at the time of inception to potentially receive a *net credit* instead of a debit premium paid or zero cost to own the structure.

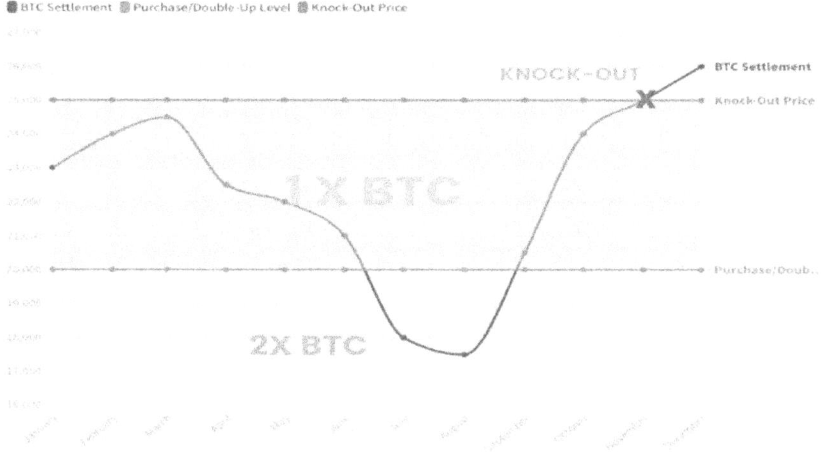


[lxxi]

The purchase/double-up barrier level in the current Accumulator example is a 2x multiplier, meaning that this structure allows for the purchase of 1 bitcoin at $20,000 during each monthly observation period through expiration in exchange for the obligation to buy two bitcoin per month at $20,000 were the price of bitcoin to settle below $20,000 during any given monthly observation.

The final customization aspect of this structure is the knock-out barrier level, which in this example is set to $25,000. For each monthly observation period where the reference price for BTC/USD is within the range of $20,000 and $25,000, the structured product owner buys one bitcoin at $20,000. If the BTC/USD reference rate settles above the knock-out barrier level set at $25,000 during any observation period, then no bitcoin is purchased for that observation period. For this current Accumulator example, the structure terminates and expires were the knock-out barrier level to be breached during any observation period. It is possible to build an Accumulator that has a knock-back-in function were the BTC/USD reference rate to settle below the knock-out level during future observation periods, reactivating the purchase/double-up level of the structure at $20,000.

## COUPON ACCUMULATOR

A Coupon Accumulator is a structured product designed to distribute a predefined yield based on the notional value of a purchase obligation assumed by the product owner. The general premise of a Coupon Accumulator is to set a purchase level for bitcoin that is below the current BTC/USD reference rate at trade inception while also receiving a yield were bitcoin prices to move higher from the trade inception reference rate. As a customizable structure, there are numerous variations to a Coupon Accumulator. Collateralized with USD, this approach to accumulation provides both a yield opportunity plus a bitcoin purchase obligation at a fixed price level.

In the following chart, we can analyze a monthly Coupon Accumulator with a 6-month duration that provides 1/6$^{th}$ of the total yield payment to the product owner when the BTC/USD settlement reference rate is above the strike price during each monthly observation period. In exchange for potential yield, the product owner assumes the obligation to buy 1/6$^{th}$ of the notional contract value of bitcoin at the purchase barrier level, were the BTC/USD reference rate to settle below the purchase level of $18,000 during any observation period. If bitcoin trades between the purchase and yield barrier levels, then no coupon payment is distributed for that observation period, and no bitcoin is purchased.

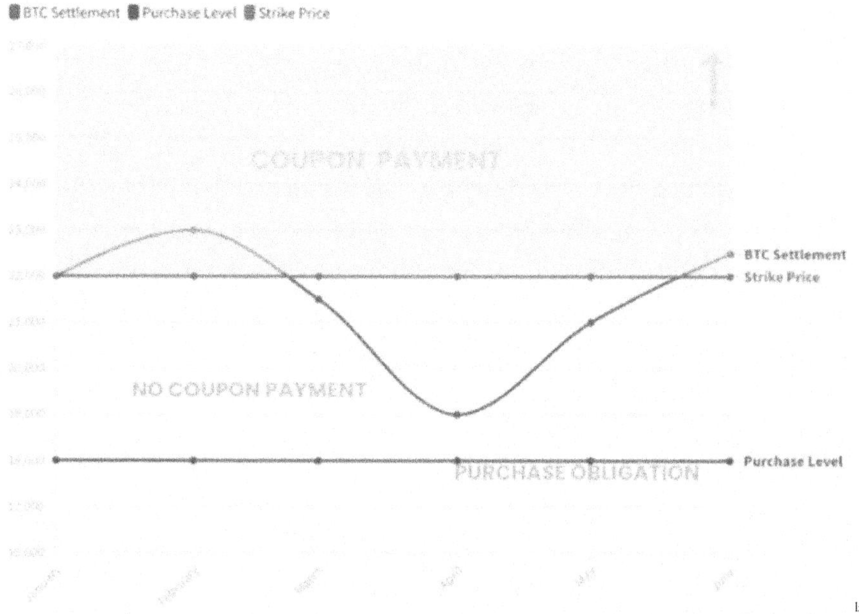


[lxxii]

For the Coupon Accumulator in focus, the optimal yield scenario is shown below. Notice during each observation period over the 6-month duration, the BTC/USD reference rate settled above the strike price, providing 1/6$^{th}$ of the net yield payment during each monthly observation as seen below:

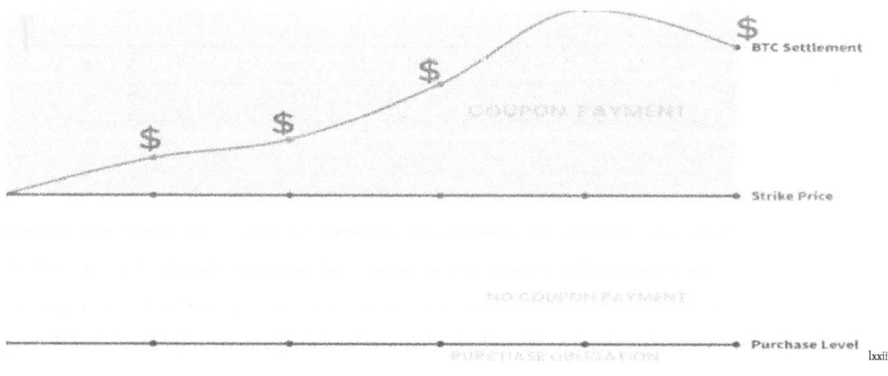


[lxxiii]

For a difference scenario analysis with the same Coupon Accumulator structure, notice that during the first monthly observation period the BTC/USD reference rate settled above the strike price, providing a coupon payment to the structured product owner. From there, the BTC/USD reference rate settled below the purchase level for the following four monthly observations, obligating the structured product owner to buy bitcoin at the purchase barrier level during each of those observation periods. Finally, during the 6$^{th}$ and final observation period for this Coupon Accumulator structured product, the BTC/USD reference rate settled above the purchase barrier level and below the strike price, meaning no coupon yield payment was distributed to the structured product owner, nor was bitcoin purchased at the purchase barrier level.

For this scenario example, the Coupon Accumulator structured product provided 1/6 coupon distributions, 4/6 purchase obligations, and, 1/6 no coupon, and no bitcoin purchase obligation during the six total monthly observation periods as defined by the structured product.

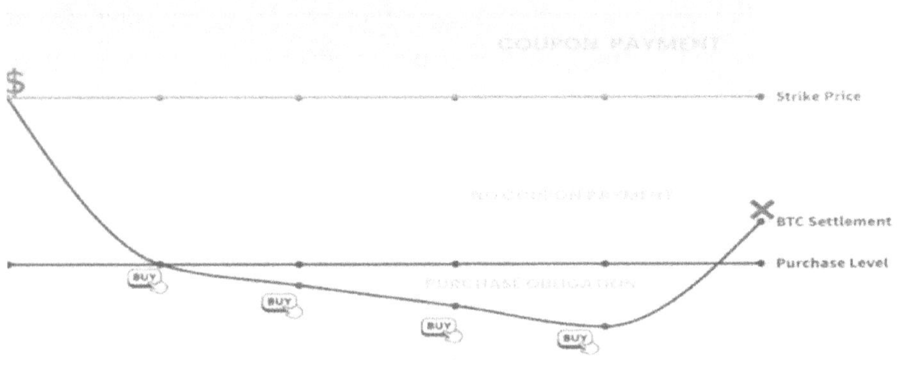


lxxiv

Building on the logic of a Coupon Accumulator, a similar yet inverse structure to the Coupon Accumulator could be developed. For example, a Coupon Decumulator could be constructed where yield is earned when the BTC/USD reference rates settles below the strike price. No coupon payment is issued when the BTC/USD reference rate settles between the strike price and the sale barrier level, and the Coupon Decumulator structure product owner has an obligation to sell bitcoin at a sale barrier level during each observation period. A Coupon Decumulator could prove attractive to market participants who are happy to sell bitcoin at the sale barrier level in exchange for the potential to earn yield.

## DECUMULATOR

Given that an Accumulator is designed for market participants seeking to purchase bitcoin at a purchase/double up barrier price level that is below the BTC/USD reference rate at trade inception of the structure, a Decumulator is assembled in a way that allows the product owner to **sell** bitcoin at a sale/double-up barrier price level that is above the BTC/USD reference rate at inception. This product can be helpful for bitcoin miners seeking to manage forward sale prices or other bitcoin holders seeking to risk mitigate conversion rates between bitcoin and the U.S. dollar.

In the following example, a weekly Decumulator has been structured to sell bitcoin at a sale/double up barrier price level of $27,000 once per week over 24 weeks. At trade inception, the BTC/USD reference rate was at $23,000. The knockout barrier was set to $21,000, allowing the structure to be owned at zero cost with no premium payment required. If the BTC/USD reference rate is within the upper sale/double-up barrier price level and the lower knockout barrier level, then the product owner will sell one bitcoin at the sale barrier level of $27,000 during each weekly observation period as defined in the product specifications.

If the BTC/USD reference rate were above the sale/double-up barrier price level of $27,000 during any weekly observation period over the life of the structure, then the product owner would sell two bitcoin at $27,000. In a scenario where the BTC/USD reference rate were to settle below the knockout level during any weekly observation period, the product owner would sell zero bitcoin at the sale/double-up barrier level, causing the structure to knock out and terminate. Upon termination or at expiration, the total quantity of bitcoin sold is path-dependent on the BTC/USD reference rate through the life of the structured product.

# BITCOIN OPTIONS

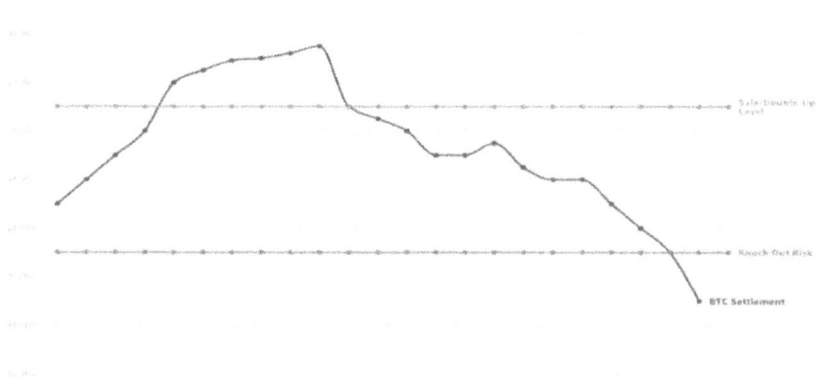


[lxxv]

Looking at the monthly Decumulator example below, which is designed to sell one bitcoin once per month for 12 months, we can analyze the quantity of bitcoin sold at the sale/double-up barrier price level of $27,000 during each observation period until expiration or a knock-out event occurs. In this monthly Decumulator example, the knock-out barrier causes the product to terminate early if the BTC/USD reference rate settles below the knock-out barrier price level of $21,000 during any observation period.

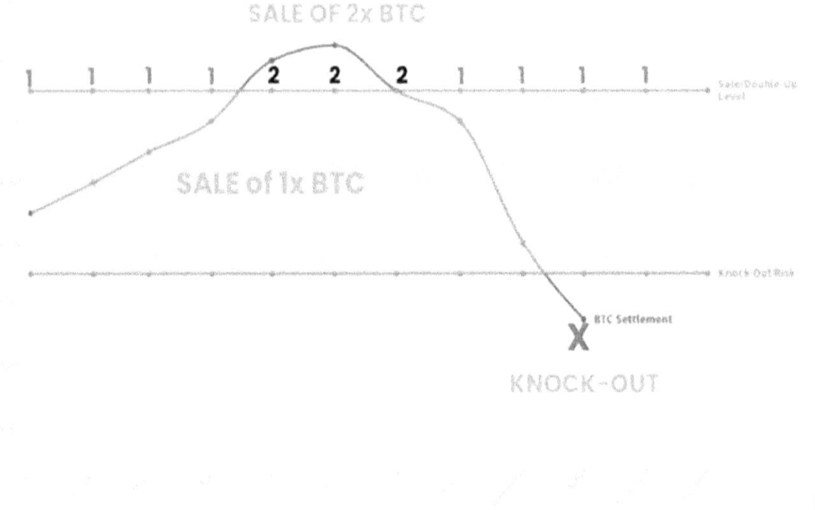


lxxvi

The results for this 12-month duration, monthly Decumulator example from month one until the knock-out barrier was breached during the month ten observation period yielded the sale of 12 bitcoin at the sale/double-up barrier price level of $27,000. Since the monthly Decumulator provides for the sale of 1 bitcoin if the reference price of BTC/USD is between the sale/double up barrier price level and the knock-out barrier price level, the chart above clearly depicts the quantity sold until the structure knocks out during the October observation, which yielded 0 bitcoin sold at $27,000 and the termination of the structure.

## NO-KO DECUMULATOR

A NO-KO Decumulator is the same structured product as a standard Decumulator except it replaces the knock-out barrier price level with a non-pricing level. When breached during an observation period, the non-pricing level does not terminate the structure, rather the structure remains active although no bitcoin is sold at the sale/double-up price level during any observation period where the BTC/USD reference rate settles below the non-pricing level. Were the BTC/USD reference rate to rise above the non-pricing level during future observation periods, then the product owner would again sell the defined quantity of bitcoin at the sale/double-up level.

For the example below, with the BTC/USD reference rate at trade inception being $27,600, a 12-week **NO-KNOCK-OUT DECUMULATOR** has been assembled that enables market participants to sell bitcoin 10% higher at $30,300 once per week when the BTC/USD reference rate is observed between the sale/double up level and the non-pricing level. If bitcoin prices trade above the sale/double up level during any given observation period during the life of the trade, then 2x bitcoin are sold at $30,300. If the BTC/USD reference rate were to settle below the non-pricing level during any observation period, no bitcoin would be sold at the sale/double-up price level.

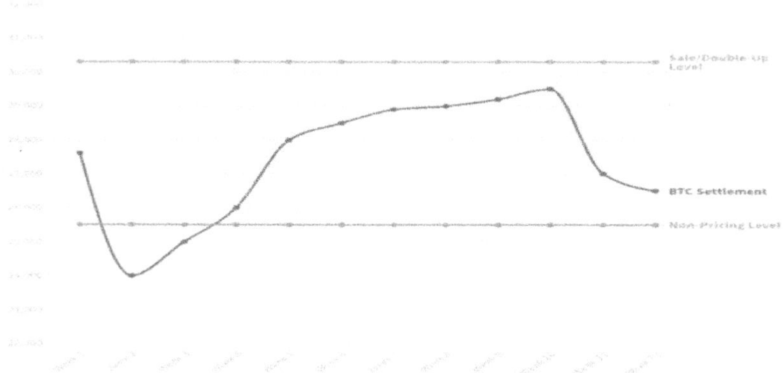


lxxvii

CHAPTER

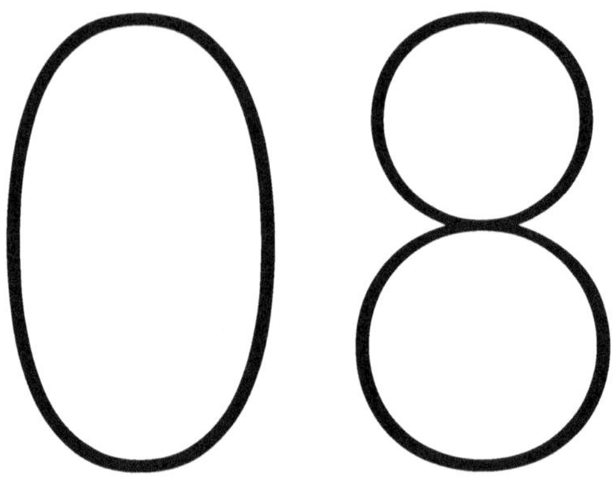

# HEDGERS ARE MADE FOR MARKETS

# CONCLUSION

The seven chapters of *Bitcoin Options: Energy Storage, Risk Management, Real Yield* encompass a practical exploration of bitcoin options markets for HODLr's, bitcoin miners, asset managers, and investors to optimize strategy and execution techniques through a risk/reward perspective. Equally so, this book acknowledges the all-encompassing interpersonal interconnectivity of the global population. Bitcoin presents an option to the world by uniting all people through an energy storage and transfer mechanism controlled by no one and accessible to everyone on the planet. Recognizing the differences between centralized and decentralized systems, and open-source systems compared to closed-source systems is essential to removing the scales from our eyes. Scales that were socially constructed by those seeking to control societal systems in darkness; removing them reveals the Light. There is a difference between the freedom to transact and restricted access to transact. Bitcoin is a digital commodity that provides a natural hedge against fraudulent hierarchies that have successfully taken control of societal "money." In practice, bitcoin is a call option against present tendencies and historical facts of fear and greed perpetuating corruption against civil society.

Once monetary tyranny takes hold within society through centralized control of the money supply, the effects force the affected population to operate within inefficient systems. Central bank mismanagement of currency supply led to an explosion of personal and national debt obligations and interest payments. This overbearing weight of debt obligation trailblazes a path toward bitcoin awareness and education as a personal energy and storage transfer mechanism, allowing people to opt out of perpetual debt servitude.

Bitcoin is peer-to-peer electronic cash with a fixed supply that solves the "double spend" accounting and computer science problems. Deployed anonymously into the public sphere, bitcoin is a decentralized protocol for tracking the storage and transfer of energy between people. This means money without centralized architecture is accessible equally to everyone.

Once the bitcoin option is realized, the goal is to acquire and hold bitcoin strategically to withstand fear and greed tendencies. As outlined in this book, options contracts can be used as a tool to incorporate bitcoin into a portfolio of assets as a hedge against instinctual power grabs. Bitcoin solves people's worldwide problem, where authoritative entities have oversupplied human society with debt while simultaneously debasing the value of prior money earned.

Money is a human concept designed to facilitate the transfer of human energy. The instrument chosen to represent "money" has historically been denominated in physical commodities, like gold and silver, deemed valuable by all people. Without value consensus, there is no interconnecting value transfer mechanism. From seashells to gold and now paper fiat currency, centralized control over these human energy storage and transfer mechanisms has proven to lead toward social systems wrought with inequality, inefficiency, and low vibration from debt burdens.

Despite the negative externalities associated with centralized control of monetary systems, the current debt-based fiat currency structure has allowed for noteworthy progress in key economic indicators like child mortality rates, increased life expectancy, expanded access to clean water, indoor plumbing, electricity, and education. While systemic debt can result in low vibrational activity, innovation often involves using debt, where a person or company with an idea needs to borrow energy in the form of money to create goods and services that provide positive life-changing solutions for people. However, such debt structures are optimized in a peer-to-peer format, not when debt is forced upon society, as we see today. If a person or company accumulates excess energy, they are free to lend that energy to others. By conducting a proper risk/reward analysis and cash flow projections based on rational assumptions for future output, the lending and borrowing of money can lead to incredible production and innovation.

Since debt comes with the obligation of repayment at interest, socializing the consequences of lending and borrowing activities unfairly burdens those with no decision-making authority on said obligations. Herein lies the core problem of centralized monetary policy. Central banks, created by artificial and unnatural laws, are private corporations, and they are incentivized to lend money to governments, who then tax people to service those debt obligations. When personal, corporate, and government debt obligations are socialized, the consequences become socialized through inflation and a tiered system for accessing credit. Central banks are incentivized to trap society in perpetual debt servitude. This is why we must free ourselves from central banking control. The process of understanding bitcoin shines light on the path towards freedom.

Historically, the most common application of socialized debt was used to perpetuate warfare by kings and queens who came to power via bloodlines and conquest. The decision-making process was inherently centralized, though the consequences were placed on the common people who would fight these wars and finance these conflicts through labor, taxation, and natural resources. When decision-makers do not feel consequences, poor choices are inevitable. Today is no different than the past, where debt is used to fund perpetual warfare and ideological conquests via education through indoctrination. The United States spends $1 trillion annually on defense to have a military presence on all continents and in over 150 countries. This is only possible through the socialization of debt.

Monetary mismanagement creates unequal barriers for people – the ever-present divide between the "haves" and "have nots." The powerless pay for the powerful via labor and taxation, which sets a foundation of unequal barriers to entering the economic system and participating in the ebb and flow of human energy. This unequal foundation sets the stage for authoritative processes that delve into a cycle of energy theft through inflation, the stagnation of work wages that cannot keep up the cost of essential survival goods and services, and the constant threat of asset seizure and imprisonment for anyone seeking to opt out of the system.

**Referring to Chapter 1, the total US public debt was observed to be around 33 trillion USD.** This number is surreal. US government interest payments were also observed at around 1 trillion USD **annually.** Will this debt burden spark a renaissance age of prosperity in the United States? No. By nearly all measurements, the citizens of the United States have not realized proportional gains to their standard of living relative to the debt burdens that they and their children and their children's children have been selected to bear. From 1971 to the present, amplified at the turn of the millennium, United States citizens were chosen as the human energy to finance perpetual global warfare and destabilization. These same people have also become subject to an expanding list of governmental agencies and a tightening grip on personal liberties through surveillance, gatekeeping, sanctions, limits, and executive orders.

Those eager to seek truth often find perplexing questions around the significant paradoxes of human life when examining peace and war, love and hate, justice, and injustice. The natural state of all people is peace, love, and the realization that we are all one. As one people, one equals one. Man, woman, black, white, purple; we are all equally one on this planet in our natural state. The common denominator for the negatives, war, hatred, injustice, and inequality, share a common theme rooted in energy transfer incentives. Peace is free for all who seek it, whereas war must be financed. Love is free for all who seek it, whereas hate must be constructed. Equality is the natural state, whereas inequality must be devised. Efficiency is harmony and justice, whereas inefficiency is imbalance and injustice.

Money represents human energy storage and transfer, and the people who control the money control the incentive structure. To hedge centralization, bitcoin represents money as a digital commodity existing within a decentralized network open to everyone and equally accessible to all. As electronic cash, bitcoin provides a bridge from the mechanical age to the electric age with peer-to-peer energy transfer that is immutable, censorship-resistant, and unconfiscatable, given the lack of need for a third-party gatekeeper.

Energy can neither be created nor destroyed, only transferred. Energy is the ability to do work. Work is the transfer of heat. Bitcoin is a heat transfer system that provides a public ledger of energy transfer transactional data blocks linked together in a perpetual chain. The laws of thermodynamics provide the foundational concept of money and the ability to represent energy storage and transfer for people on Earth through a consensus mechanism. If a person conceives and creates something of value, they shall be able to transfer that value to another who demands it in exchange for potential energy. That potential energy is money that can be stored and transferred, creating an interconnected system that incentivizes valuable creation through work.

When fraudulent systems capture money, they incentivize people to work inefficiently toward an uncommon goal. Centralization of money creation becomes a power lever sought after by those with the most pressing urge for unnatural and unequal power dynamics. Once bitcoin is acknowledged as an instrument that can usher in a new era of efficient facilitation of human energy storage and transfer, managing the instrument comes into focus. **Not your keys, not your bitcoin** is the first stage of bitcoin risk management. The characteristic of self-custody makes bitcoin a digital commodity that is unconfiscatable and censorship-resistant. From there, other risk management strategies depend on counterparty risk (centralized or decentralized) to provide market access services into spot bitcoin liquidity, futures contracts, and options contracts.

Energy Storage, Risk Management, Real Yield.

There is a saying that the fear of loss is greater than the prospect of gain. Although not valid for everyone, this saying exposes the root of survival instincts that affect most people's decision-making - especially with personal asset management. To facilitate action, risk management techniques seek to blunt instinctual force. As learned throughout this book, options contracts, futures contracts, exotic options, and structured products can optimize execution when acquiring bitcoin, selling bitcoin, seeking passive income, or managing exchange rates between other assets.

Financial instruments are often seen as tools for speculation, a market side that garners attention from mainstream media and social interaction. However, with the proper education, these instruments become essential tools for Bitcoin HODLr's, miners, asset managers, and investors. They enable the definition and implementation of controlled execution strategies, identification of real yield opportunities, and deployment of risk management tactics. This knowledge empowers you to make informed financial decisions.

Fix the money, free the internet.

Peace.

# GRAPH CITATIONS

[i]Federal Reserve of St. Louis. (2022). Federal debt: Total public debt. US Department of the Fiscal Treasury Fiscal Service. Retrieved from https://fred.stlouisfed.org

[ii]Federal Reserve of St. Louis. (2022). Government current expenditures: Interest payment. US Department of the Fiscal Treasury Fiscal Service. Retrieved from https://fred.stlouisfed.org

[iii]Pew Research Center. (2022). Average hourly wages in the United States, adjusted. Retrieved from https://www.pewresearch.org

[iv]Federal Reserve of St. Louis. (2022). M2. US Department of the Fiscal Treasury Fiscal Service. Retrieved from https://fred.stlouisfed.org

[v]Federal Reserve of St. Louis. (2022). Share of total net worth held by the top 1% (99th to 100th wealth percentile). US Department of the Fiscal Treasury Fiscal Service. Retrieved from https://fred.stlouisfed.org

[vi]Federal Reserve of St. Louis. (2022). Share of total net worth held by the top 0.1% (99th to 100th wealth percentile). US Department of the Fiscal Treasury Fiscal Service. Retrieved from https://fred.stlouisfed.org

[vii]GoldPrice.org. (2022). How much gold. Retrieved from https://www.gold.org/goldhub/data/how-much-gold

[viii]World Gold Council. (2022). Retrieved from https://www.gold.org

[ix]US Geological Survey. (2022). How much silver has been found worldwide? Retrieved from https://www.usgs.gov/faqs/how-much-silver-has-been-found-world?qt-news_science_products=0#qt-news_science_products

[x] SilverPrice.org. (2022). Retrieved from https://www.silverprice.org

[xi] Bitcoin Concepts. (2022). Bitcoin supply mined per year. Retrieved from https://www.bitcoinconcepts.org

[xii]Federal Reserve of St. Louis. (2022). M2 total dollar supply. US Department of the Fiscal Treasury Fiscal Service. Retrieved from https://fred.stlouisfed.org

[xiii]CoinMetrics. (2022). BTC daily issuance, supply, halvings. Retrieved from https://www.coinmetrics.io

[xiv]Blockchain.com. (2022). Total circulating bitcoin. Retrieved from https://www.blockchain.com

[xv]Aristotle. (1944). *Nicomachean ethics, Book I, Section 1259a.* In Aristotle in 23 Volumes, Vol. 21 (H. Rackham, Trans.). Harvard University Press. Retrieved from http://www.perseus.tufts.edu/hopper/text?doc=Perseus%3Atext%3A1999.01.0058%3Abook%3D1%3Asection%3D1259a

[xvi]Poitras, G. (2008). The early history of option contracts. Simon Fraser University. Retrieved from https://www.sfu.ca/~poitras/heinz_$$.pdf

[xvii]Poitras, G. (2008). The early history of option contracts. Simon Fraser University. Retrieved from https://www.sfu.ca/~poitras/heinz_$$.pdf
[xviii]Black, F., & Scholes, M. (1973). The pricing of options and corporate liabilities. *Journal of Political Economy, 81*(3), 637-654. Retrieved from https://www.cs.princeton.edu/courses/archive/fall09/cos323/papers/black_scholes73.pdf
[xix]New Liberty Standard. (n.d.). Bitcoin exchange rate. Retrieved from https://newlibertystandard.io/exchange
[xx]Laevitas. (2023). Total crypto options notional volume as % of total crypto derivatives market. Retrieved from https://www.laevitas.ch
[xxi] Divine, J. (2023). Original graph. [Bitcoin to Argentine Peso Graph].
[xxii] Divine, J. (2023). Forward curve: Aug 2022 – July 2023. [BTC to Turkish Lira Graph]
[xxiii]Bilello, C. (2023). Global currencies: 10-year return versus the dollar. XE.com. Retrieved from https://www.xe.com
[xxiv]Divine, J. (2023). Original graph. [Calls & Puts Graph].
[xxv] Divine, J. (2023). Original data. [Call Options Data Sheet].
[xxvi]Divine, J. (2023). Forward curve: Aug 2022 – July 2023.
[xxvii]Divine, J. (2023). Bitcoin Forward curve: Aug 2022 – July 2023.
[xxviii]Divine, J. (2023). Intrinsic & Extrinsic Value: Aug 2022 – July 2023.
[xxix]Zimmerman, M. J., & Bradley, B. (2019). Intrinsic vs. extrinsic value. In E. N. Zalta (Ed.), The Stanford Encyclopedia of Philosophy (Spring 2019 Edition). Retrieved from https://plato.stanford.edu/archives/spr2019/entries/value-intrinsic-extrinsic/
[xxx]Cambridge Dictionary. (n.d.). Extrinsic value. Retrieved from https://dictionary.cambridge.org/us/dictionary/english/extrinsic-value
[xxxi]Divine, J. (2022). BTC/USD August 26, 2022, options market.
[xxxii]Divine, J. (2022). BTC/USD Aug 26th, 2022, options market.
[xxxiii]Simpler Trading. (n.d.). How theta decay works. Retrieved from https://www.simplertrading.com/blog/how-theta-decay-works
[xxxiv]Divine, J. (2022). Price discovery range, BTC/USD August 26, 2022.
[xxxv]Milken Institute. (n.d.). The Greeks. Retrieved from https://www.5minutefinance.org/concepts/the-greeks
[xxxvi]Divine, J. (2023). Volatility graph, lognormal and normal.
[xxxvii]Divine, J. (2023). Standard deviation graph.
[xxxviii]Career Foundry, Ejaz, N. (2023). Standard error vs. standard deviation: What's the difference? Retrieved from https://careerfoundry.com/en/blog/data-analytics/standard-error-vs-standard-deviation/
[xxxix]Divine, J. (2023). Realized volatility graph.
[xl]Divine, J. (2022). Calculating realized volatility, BTC/USD August 1, 2021 - August 1, 2022.

[xli]CME Group. (n.d.). Introduction to options volatility. Retrieved from https://www.cmegroup.com/education/courses/introduction-to-options/discover-options-volatility.html

[xlii]Divine, J. (2022). Implied volatility data chart, August 26, 2022.

[xliii]CME Group. (2023). Volatility calculation graph. Retrieved from https://www.cmegroup.com

[xliv]Divine, J. (2022). BTC August 26th, 2022 2300 call.

[xlv]CME Group. (2023). IV. Retrieved from https://www.cmegroup.com

[xlvi]Divine, J. (2022). BTC August 26th, 2022 2300 call.

[xlvii]Divine, J. (2022). Call option analysis.

[xlviii]Divine, J. (2022). 2300 options Greeks.

[xlix]Simpler Trading. (n.d.). How theta decay works. Retrieved from https://www.simplertrading.com/blog/how-theta-decay-works

[l]Career Foundry, Ejaz, N. (2023). Standard error vs. standard deviation: What's the difference? Retrieved from https://careerfoundry.com/en/blog/data-analytics/standard-error-vs-standard-deviation/

[li]The Block. (n.d.). Annualized BTC volatility 30d. Retrieved from https://www.theblock.co/data/crypto-markets/prices/annualized-btc-volatility-30d

[lii]CoinMarketCap. (n.d.). Retrieved from https://www.coinmarketcap.com

[liii]CoinMarketCap. (n.d.). Retrieved from https://www.coinmarketcap.com

[liv]CoinMarketCap. (n.d.). Retrieved from https://www.coinmarketcap.com

[lv]Sfeir, E. (2023). Options and structured products. Blockfills.

[lvi]TradingView. (2023). Etherium to Bitcoin. Retrieved from https://www.tradingview.com

[lvii]ExxaBlock. (2021). [Data report].

[lviii]ExxaBlock. (2021). [Data report].

[lix]ExxaBlock. (2021). [Data report].

[lx]Amberdata. (2023). [Data report].

[lxi]ExxaBlock. (2021). [Data report].

[lxii]Amberdata. (2023). [Data report].

[lxiii]Laevitas. (2023). [Data report].

[lxiv]CME Group. (2023). Commitment of traders (COT) tool. Retrieved from https://www.cmegroup.com/tools-information/quikstrike/commitment-of-traders.html

[lxv]ICE. (2023). Retrieved from https://www.theice.com

[lxvi]Investing.com. (2023). Retrieved from https://www.investing.com

[lxvii]Investing.com. (2023). Retrieved from https://www.investing.com

[lxviii]Ortiz, T. (2023). TPAC. [Data report].

[lxix]Ortiz, T. (2023). TPAC. [Data report].

[lxx]Ortiz, T. (2023). TPAC. [Data report].

[lxxi]Ortiz, T. (2023). TPAC. [Data report].
[lxxii]Ortiz, T. (2023). TPAC. [Data report].
[lxxiii]Ortiz, T. (2023). TPAC. [Data report].
[lxxiv]Ortiz, T. (2023). TPAC. [Data report].
[lxxv]Ortiz, T. (2023). TPAC. [Data report].
[lxxvi]Ortiz, T. (2023). TPAC. [Data report].
[lxxvii]Ortiz, T. (2023). TPAC. [Data report].

www.ingramcontent.com/pod-product-compliance
Lightning Source LLC
LaVergne TN
LVHW010650110826
845149LV00014B/3019

* 9 7 9 8 9 8 9 9 6 8 4 0 4 *